Aus dem Meteorologischen Institut der Universität Hamburg
(Direktor: Prof. Dr. K. Brocks)

Hamburger Geophysikalische Einzelschriften
Herausgegeben von den Geophysikalischen Instituten
(Fachgebiete: Meteorologie, Ozeanographie, Physik des Erdkörpers)
der Universität Hamburg

Heft 9

Die Mesostruktur des Windfeldes an der Grenze zwischen Troposphäre und Stratosphäre

Mesostructure of the Wind Field at the Boundary between Troposphere and Stratosphere

von

Georg Duensing

Hamburg 1967
Cram, de Gruyter u. Co

Preis: DM 14,—

Diese Arbeit wurde als Dissertation zur Erlangung des Doktorgrades
von der Mathematisch-Naturwissenschaftlichen Fakultät
der Universität Hamburg angenommen.

I n h a l t s v e r z e i c h n i s / C O N T E N T S

Zusammenfassung

Die Arbeit benutzt Winddaten, die in den Jahren 1962 und 1963 auf 143 Flügen mit einer Düsenverkehrsmaschine der Deutschen Lufthansa vom Typ Boeing 707 - 430 registriert wurden. Der mittlere Fehler der einzelnen Windmessung ist bei Benutzung des Doppler-Radar-Gerätes kleiner als 2,9 m sec^{-1}.

Die mittlere Fluktuation des Windes, definiert als die Wurzel aus der mittleren Summe der quadrierten vektoriellen Differenzen zweier benachbarter Windwerte im Abstand von etwa 100 km, beträgt 6 m sec^{-1}. Dagegen liegt die mittlere Streuung des Windes, gebildet aus den Quadraten der vektoriellen Differenzen zwischen aktuellen Werten und den über 200 km geglätteten Mitteln, bei 2 m sec^{-1}. Die Streuung des Windfeldes ist in der Mesostruktur horizontal-isotrop sowie im Bereich der Strahlströme und an der Tropopause am größten. Die höchsten Wind-Gradienten (über 40 m sec^{-1} auf 100 km) werden am Rande des Strahlstromes beim Einflug in Höhentröge festgestellt.

Für den "erflogenen" Vergleich der flug-geostrophischen und aktuellen Querwind-Komponenten beträgt der Korrelationsfaktor 0,73 (Zufallshöchstwert 0,211). Der Vergleich der aktuell-erflogenen Winde mit geostrophischen Winden aus der Wetterkarte zeigt Korrelationsfaktoren über 0,7. Die mittlere Streuung der vektoriellen Differenzen zwischen erflogenen und geostrophischen Winden aus der Wetterkarte beträgt 15,5 m sec^{-1}. Dieser hohe Wert ergibt sich durch die ungenaue Kenntnis über die Lage des jet stream und der Höhenzyklonen. Der mittlere Fehler des geostrophischen Windes aus der Wetterkarte liegt heute zwischen 7 m sec^{-1} und 10 m sec^{-1} je nach Position zu den aerologischen Stationen. Der mittlere Windfehler der Wetterkarte wird sich auf 5 m sec^{-1} verringern, falls Wind-Messungen mit Flugzeugen zusätzlich zu den Wind-Messungen mit Radiosonden verwendet werden. Dem geostrophischen Wind, wie er der Wetterkarte des 200-mb- oder 300-mb-Niveaus (Termin 00 GMT oder 12 GMT) entnommen wird, ist nachts ein zusätzlicher Nordwind und tags ein zusätzlicher Südwind zu addieren, um den erflogenen Wind zu erhalten.

SUMMARY

The data of winds used in this paper were measured in 1962 and 1963 during 143 Atlantic-crossings with Boeing 707-430 of the Deutsche Lufthansa. These winds based on a Doppler-Radar-System show a accuracy higher than 2.9 m sec^{-1}.

The mean fluctuation of wind defined as the square root of the mean sum of the squares of the vector differences between two wind-data is 6 m sec^{-1}. The spacing between these data is about 100 km. The standard deviation of the wind computed from the vector differences between actual wind-data and the means which are being smoothed over a distance of 200 km is about 2 m sec^{-1}. The standard deviation of the wind and the mean fluctuation of wind are horizontaly isotropic in meso-scale and greatest at the jet stream and at the tropopause. Windshears of more than 40 m sec^{-1} per 100 km are measured near the jet stream core when the plane passes old cyclones or troughs in the free atmosphere.

The correlation coefficient of the geostrophic and the actual crosswind components, both measured by aircraft, is 0.73 which is far beyond the limit of probability. The correlation coefficient between the actual wind measured by aircraft instruments and the geostrophic wind read from weather charts is more than 0.7. The standard deviation of the vector differences between Doppler winds and geostrophic winds taken from weather charts is about 15.5 m sec^{-1}. It is of greatest importance to know the position of high level cyclones and the structure of the jet stream in order to get the best accuracy of wind. Nowadays the accuracy of geostrophic winds estimated from weather charts is between 7 m sec^{-1} and 10 m sec^{-1} depending on the distance from the aerological stations. When wind measurements by aircraft are added to the rawin data the accuracy will increase for about 5 m sec^{-1}.

The geostrophic wind taken from 200-mb- or 300-mb-weather-charts (time of observation 00 GMT, 12 GMT) systematically differs from the actual wind measured by aircraft. To get the mean actual wind one must add an ageostrophic north-wind at daytime and an ageostrophic south-wind at nighttime to the mean geostrophic wind.

1. Abstract Chapter 2 - Chapter 9

Abstract Chapter 2 (page 37 - 41)

PROBLEMS OF MEASUREMENT AND ANALYSIS OF METEOROLOGICAL DATA RECORDED BY JET-AIRCRAFT

The development of the meteorological science is sponsored to a high degree by the progresses made in the techniques of aircraft, rocket and satellite. Operational and flight plannings for supersonic aircrafts crossing the Atlantic need informations of density fields in different altitudes. It is very difficult to get accurate measurements of pressure, height above sea level, and temperature by aircraft observations. Therefore the windfield gets priority for the wind can be observed by many planes. The accuracy of Doppler-wind-data is sufficient to discuss fluctuations, variations, and discontinuities of the upper wind field. These qualities are most important for the density field. Therefore the discussed qualities of the upper windfield give first informations for supersonic flight planning.

Nowadays many aircrafts of the civil air-line operating companies cross the Atlantic. Most of the jet clippers fly between 300 mb and 200 mb. Measurements made by aircraft complete the measurements made by radiosondes. The radiosonde network gives meteorological informations in large scale (macro-structure). Isobaric and continuous measurements made by aircraft give informations in meso- and micro-scale. Therefore these observations and measurements by aircraft should be analysed by synoptic meteorologists in order to forecast upper weather charts. But the chance, given by very good organized observations and recorded data of weather conditions along the flight path is not exploited. By modern electronic computers it is possible to record data at flight. The computers are able to analyse, to compare the records, and to plot weather charts with meso-scale qualities.

The air-traffic safety control for supersonic aircraft is only possible by using modern electronic computers. Therefore it should be possible to use these constructions for recalling meteorological data recorded by aircraft. The computers of the air-traffic safety control should also be able to transmit the meteorological data to electronic computers installed at meteorological offices.

This paper shows the importance of receiving meteorological data by flight-recorder. The scale of the atmosphere has been described by Reiter (50) . We follow his example in this form:

Macro structure	more than 300 km horizontal distance
Meso structure	more than 10 km and less than 300 km
Micro structure	less than 10 km horizontal distance

Meteorological measurements by aircraft complete measurements by radiosonde in macro-scale and they give detailed informations in meso-scale . One may notice qualities of the atmospheric micro structure in bad flight conditions, disagreeable vibrations, and short period rolling of the aircraft. In aviation meteorology these disturbances are defined "Clear Air Turbulence" (CAT) .

The subdivision of the atmospheric scale (macro-, meso-, and micro structure) did result from aeronautical and not from meteorological aspects. Therefore one has to expect that the meso structure of the atmosphere could produce phenomena like CAT in jet operations.

This paper applies data recorded by Boeing 707-430 D-ABOD of Deutsche Lufthansa. The data have been recorded during 143 Atlantic-crossings in the periods March 1962 - May 1962 and March 1963 - Octobre 1963. Wind—data have been calculated for each point in intervals of 6 minutes flight-time. There are 8968 accurate wind-data. In the following the wind-data calculated from the recorded parameters are called "flight winds" as distinguished from "spot-winds" computed from the aeronautical parameters shown at the panel.

Abstract Chapter 3 (page 42 - 78)

HORIZONTAL FLUCTUATIONS OF THE UPPER WINDFIELD

3.1 REMARKS

In the following years it would be possible to forecast minimum-fuel-tracks for supersonic aircraft. Therefore it is necessary to have a good knowledge of the meso structure of the atmosphere and to forecast atmospheric qualities in mesoscale. Nowadays the atmospheric micro- and meso structure is neglected in analysing and forecasting meteorological fields. The air-particles get a strong acceleration in micro- and meso scale. Therefore in these regions one has to expect ageostrophic components of the windfield.

A uniform geostrophic windfield will be remarkably disturbed by each sufficiently strong vertical momentum, if the ascending air reaches the level of steady air flow with a horizontal velocity different from the geostrophic wind speed. In most of the time the atmosphere has no strong stationary sources of "primer momentums", which one can analyse in macro structure of the circulation. Convective perturbations of the upper windfield exist in many forms. The upper windfield will be disturbed by cumulonimbus with ascending or descending air-flow, if the wind velocity at ground is not equal to those in upper levels. Also vertical exchange-processes near the tropopause produced by radiation processes or wind shears disturb the horizontal geostrophic windfield. Therefore a uniform undisturbed geostrophic windfield cannot exist at upper levels. One has to expect strong local and temporal changes in the upper windfield.

In this paper the following definition named "mean wind fluctuation" will be used

$$f = \left(\frac{\sum_{i=1}^{n} (x_i - x_{i+1})^2}{n} \right)^{1/2}$$

n = number of observations

x_1 = value of measurement at position i

x_{i+1} = value of measurement at position i+1, 6 minutes of flight-time after position i

3.2 ACTUAL AND GEOSTROPHIC CROSS WIND

The radioaltimetry gives the "flight geostrophic crosswind" (so called in this paper to distinguish the "geostrophic crosswind" or geostrophic wind taken out of weather charts). The Doppler-radar measures the actual crosswind. These two independent crosswind-data may be correlated. Therefore the recorded parameters have been integrated by planimeters. The time-intervals have been fixed by readings of pressure and absolute altitude registered in flight logs. From these mean values one can define a mean actual and mean flight geostrophic crosswind for each interval.

Figure 1 (page 49) shows the correlation of flight-geostrophic and actual crosswind-components both integrated over an interval of more than 15 minutes. The correlation coefficient is 0.73 , much above a fortuitous result. This high correlation of actual and flight geostrophic crosswind-data in lower stratosphere and upper troposphere near the tropopause suggests to examine the data with regard to different weather situations.

Figure 2 (page 49) shows a diagram of the crosswind-components and temperature analysed by records made on a flight of Deutsche Lufthansa on 6.4.1962 (flight level 214 mb), crossing the tropopause at 35° West (see chapter 7.1).

The accuracy of a single measurement permits only a statistical analysis of flight-geostrophic data. As there exist instrument-errors which are described in chapter 9 it is necessary to integrate the parameters over an interval of 10 minutes of flight-time. In this case the standard deviation of a single measurement is about $10\ \mathrm{m\ sec^{-1}}$. The standard deviation reduces to $5\ \mathrm{m\ sec^{-1}}$ at a flight path about 300 km. The errors of measurement decrease, if no change of barometric altitude will happen.

3.3 THE MEAN UPPER WINDFIELD AND ITS MESO-STRUCTURE

The meso structure of the atmosphere has been analysed in the following systems:

System 1 : Co-ordinates, axis geographic north and east

System 2 : Co-ordinates, axis parallel and perpendicular to heading

System 3 : Co-ordinates, axis parallel and perpendicular to streamlines of the windfield

The standard deviation of a single wind measurement is less than 2.9 m sec^{-1} (see chapter 9).

The mean wind fluctuation (definition page 10, page 45), calculated from differences between two neighbouring flight-winds (distance 6 minutes of flight-time) is 5.8 m sec^{-1}.

The mean wind fluctuation (calculated from system 3) perpendicular to the actual wind direction is smaller than those parallel to the actual wind direction.

	parallel	perpendicular
1962	4.2	3.9
1963	4.7	4.4 m sec^{-1}

These results are not equivalent to the findings of Kao and Woods (29) who analysed flights of Project "Jet Stream" in USA.

The wind velocities measured by an aircraft flying parallel and perpendicular to jet streams (Project Jet Stream 1956 - 1957) have been analysed by S.K. Kao and H.D. Woods. It may be noted that in their paper the variance of the horizontal velocities shows that, in the case of flights parallel to the flow, the variance of the transversal velocity is almost twice as that of the longitudinal velocity. In the case of the flights perpendicular to the flow, the variances of the longitudinal and transversal components are of the same magnitude. The flights of project Jet Stream have been navigated parallel or perpendicular to the axis of the jet stream. It may be that the aircraft did not exactly follow a streamline. In such a case the aircraft may have recorded high differences, flying in an area with strong wind gradients (higher wind speed near the core of the jet stream and lower wind speed beyond it).

In order to show the characteristics of meso scale fluctuations in the jet stream and in the atmosphere, flights have been selected in groups which were either parallel (flights with headwind or tailwind) or perpendicular (flights with wind at the right or at the left side of the track) to the mean flow. In the analysis, the wind velocities have been split into two components, parallel and perpendicular to the mean flow. The recorded data have been smoothed and averaged by equation 3.3.1 - 3.3.4 (page 53). Table 1 gives the results of these testings. The high mean wind fluctuation shows a strong variability of wind in meso-scale (100 - 200 km). In most cases the tendency (that is the increase or decrease of wind velocity on the track) of the large-scale windfield is constant. The recorded material shows a superposition of small-scale fluctuations (microstructure) to the mesostructure of the atmosphere. The mean fluctuation in micro-scale is smaller than 1 m sec^{-1}.

The mean flight wind vector is smaller than the mean geostrophic and the mean cyclostrophic wind vector (calculated from an upper level weather chart). In most instances a strong gradient of pressure between two neighbouring stations will be equalized and not concentrated in a small region. But if there exists a small region with high wind speed and strong pressure gradient close to a region with low pressure gradient, the flight wind will be smaller than the geostrophic wind calculated from analysed weather charts.

3.4 THE MESO-STRUCTURE OF THE UPPER WINDFIELD IN RELATION TO THE WIND DIRECTION

Figure 3 (page 59) shows the dependence of the mean wind fluctuation on the wind direction. The abscissa shows the wind direction, the ordinate shows the standard deviation of the difference vectors (geostrophic wind - flight wind) and the mean wind fluctuation. Values are given in m sec^{-1}. The curves above give the standard deviations of the difference vectors flight wind - geostrophic winds.

In this diagram the curves below present the mean wind fluctuations for the components being longitudinal and transversal to the actual flow. The neighbouring wind data have a distance of about 100 km (6 minutes of flight time).

The mean wind fluctuation is about 4 m sec^{-1}. For winds from the southeast and the southwest the mean wind fluctuation of the longitudinal component is more than 0.5 m sec^{-1} above that of the transversal component. For winds from the south and first of all from northwest to northeast the collectives present a mean wind fluctuation for the component transversal to the actual wind direction a little bit higher than that of the component longitudinal to the actual wind direction. It may be that the air particles are flowing on stream lines shaped like ellipses. These results corroborate the synoptic knowledge that in temperate zones easterly winds have less persistence than westerly winds.

The great differences between geostrophic winds and flight winds shown in figure 3 (page 59) for northeastwinds resulted from errors in analysing weather situations. In our case an error could not be noticed for a few days, because an easterly wind stronger than 30 m sec^{-1} came into existence over the Northatlantic. Disadvantageously no radiowind observation measured easterly wind so that westerly winds are analysed. The standard deviation of the longitudinal ageostrophic component is much higher than those of the transversal ageostrophic component. Remarkable is the low standard deviation for the collectives with southwinds. These low data may be realized by more observations of the upper atmosphere generally. The limit of 4 m sec^{-1} is given by the mean wind fluctuation. Mainly in collectives of westwinds and southwestwinds, differences between geostrophic winds and flight winds are produced by errors in analysing the jet stream core. The accurate position of the jet stream core is unknown because of the large distances of the radiosonde-stations.

3.5 THE MESO-STRUCTURE OF THE UPPER WINDFIELD IN RELATION TO THE WIND VELOCITY

With increasing wind velocity the mean wind fluctuation increases only a little.

mean wind fluctuation	wind velocity
3.5 m sec^{-1}	5 m sec^{-1}
5.5 m sec^{-1}	50 m sec^{-1}

The atmospheric turbulence differs from the pattern of "technical turbulence" by a large horizontal and small vertical extension, rotation of the atmospheric system, and weatherlike phenomena. The turbulences of the free atmosphere are produced by wind shear, convection, or orography. Generally, convective perturbations have a smaller effect than orographic obstacles at equal heights, because the surrounding horizontal airflow accelerates a convective ascending air particle. Only the difference of horizontal velocities between the undisturbed horizontal airflow and the ascending air is active for the momentum of perturbation. Therefore the increase of the mean wind fluctuation in connection with windspeed is smaller than that in the frictional layer near the ground. Also the method of measurement and the time between two aircraft wind measurements (distance) is important for the increase of the mean wind fluctuation.

The vectorial differences (see figure 4, page 65) between geostrophic winds (computed from weather charts) and actual flight winds have been calculated in coordinates lined with geostrophic flow and correlated with the velocity of the flight wind. The standard deviation of the longitudinal component of the difference-vector (between geostrophic and flight wind) is greater than 1 m sec^{-1} above that of the cross-component. The data of the standard deviation of the difference-vectors (longitudinal and transversal to the mean wind-vector) are shown in figure 4 (page 65):

the longitudinal component of difference-vector :

windspeed	standard deviation about
less than 30 m sec^{-1}	10 m sec^{-1}
above 40 m sec^{-1}	25 m sec^{-1}

the transversal component of difference-vector :

windspeed	standard deviation about
less than 5 m sec^{-1}	3 m sec^{-1}
35 m sec^{-1}	9 m sec^{-1}

3.6 THE MESO-STRUCTURE OF THE UPPER WINDFIELD IN RELATION TO THE TROPOPAUSE AND THE JET STREAM

3.6.1 DEFINITIONS

The recorded data have been discussed in dependence on the positions to the jet stream and to the tropopause. The aircraft measurements shall be ordered in a s-formed stream. In this stream from north to south there exist the polar and subtropic jet stream, in the middle there may be a jet stream which unites both cores. Every datum shall be either tropospheric or stratospheric. Therefore it must be accepted that there exists no multiple tropopause or several discrete, overlapping "leaves".

The discussion of the data in reference to the prevailing type of jet stream gives no significant differences.

This paper does not consider the WMO-definition of jet stream. WMO defined an arbitrary 30 m sec^{-1} limit in wind speed along the jet stream axis. Also the defined horizontal and vertical gradients of wind speed are neglected. In this paper the band with higher speed has been defined as axis of the jet stream.

In the following, three scales are defined:

Scale 1 (vertical-scale):

0	no association possible	
	Tropospheric flight:	
1	flight-level more than 1000 m	below tropopause
2	flight-level from 1000 m to 500 m	below tropopause
3	flight-level from 500 m to 200 m	below tropopause
4	flight-level from 200 m to 0 m	below tropopause
	Stratospheric flight:	
5	flight-level from 0 m to 200 m	above tropopause
6	flight-level from 200 m to 500 m	above tropopause
7	flight-level from 500 m to 1000 m	above tropopause
8	flight-level more than 1000 m	above tropopause
9	collective for the control of data	

The height of the tropopause has been read from the tropopause-chart of the aviation weather station Frankfurt. These charts have been analysed by use of radiosonde-measurements exclusively. The data have been collected to collective 4 or 5 according to the air-temperature recorded by aircraft in consideration of all other atmospheric parameters. The data taken from the tropopause-chart have been linearly interpolated.

The horizontal association of the data followed the 200 mb analyse. Scale 2 is orientated cross to the jet stream core, scale 3 is orientated parallel to the jet stream core .

Scale 2 (cross-scale):

0 no association possible

1 left of
2 in the polare jet stream core
3 right of

4 between polare and subtropic jet stream

5 left of
6 in the subtropic jet stream core
7 right of

8 in the jet stream core, polare and subtropic jet stream united

9 collective for the control of data

Scale 3 (parallel-scale):

0 no association possible

1 beginning
2 centre of cyclonic current
3 end

4 point of inflection cyclonic - anticyclonic

5 beginning
6 centre of anticyclonic current
7 end

8 point of inflection anticyclonic - cyclonic

9 collective for the control of data

The figures of chapter 3.6 show in the ordinate the number of the collective defined above. The abscissa shows the wind speed or the mean wind-fluctuations. In figure 6, 7, 8 (page 72, 74) the two upper diagrams present the mean wind-fluctuation (and standard deviation of vector differences between geostrophic wind and flight wind) of the longitudinal and transversal component of the vector-difference, the diagram below presents the absolute value of modulus of vectorial difference.

3.6.2 STATISTICAL ANALYSIS OF THE FLUCTUATIONS AND AGEOSTROPHIC COMPONENTS OF THE UPPER WINDFIELD IN THE "VERTICAL SCALE"

Figure 5 (page 70) shows the mean flight wind velocities in dependence on the distance from the tropopause. In the troposphere the northern flight wind component is higher than in the stratosphere. Above the tropopause a stronger southerly flight wind component is prevailing. The maximum of the westerly flight wind component has been found directly above the tropopause. The maximum of the scalar wind velocity has been found below the tropopause. The high mean scalar wind speed continued more into the stratosphere than into the troposphere.

Figure 6 (page 72) shows the mean wind fluctuation. There is usually a strong dependence of fluctuation on the distance from the tropopause. The functional relationship varies with the height. The maximum of the mean wind fluctuation is at the tropopause. Near the tropopause the mean wind fluctuation of the component transversal to the actual wind direction is 6 m sec^{-1}. It is about 2 m sec^{-1} higher than those of the upper troposphere and of the lower stratosphere. The component longitudinal to the actual wind direction increases to about 0.5 msec^{-1} above the tropopause. Near the tropopause wind differences may result from cross circulations in the macro-scale air flow. As it seems, the upper windfield in the large scale, defined by the macro-scale field of pressure, is not disturbed by the effects of meso structure. The primary windfield must be change in large scale by the meso-scale cross circulations, if the "perturbations" work in the same direction over a long time.

The thick curves (figure 6, page 72) show the standard deviation of the vectorial differences (geostrophic wind - actual flight wind). The standard deviation of the differences between cyclostrophic wind and actual flight wind is marked by fullstop. The maximum of the standard deviation is near the tropopause. The data of the cyclostrophic wind are more uncertain. The inaccurate knowledge about the position of the jet stream core and the uncertainty about the tight beam of the field with high windspeed result in larger differences of the longitudinal component than those of the transversal component of the vectorial differences between actual flight winds and geostrophic winds.

3.6.3 STATISTICAL ANALYSIS OF THE FLUCTUATIONS AND AGEOSTROPHIC COMPONENTS OF THE UPPER WINDFIELD IN THE "CROSS-SCALE"

The cross-scale is orientated normal to the axis of the jet stream (figure 7, page 74). The maxima of the standard deviation (vector differences between flight wind and geostrophic wind) have been found in the jet stream cores. But the mean wind fluctuations of the components tangential and normal to the actual wind direction are not equal in behaviour.

The mean wind fluctuation of the component normal to the actual wind speed is greater:

1 Subtropic jet stream core, collective 6
2 polar jet stream core, collective 2
3 polar jet stream, anticyclonical shearing (on the right of the core), collective 3

The mean wind fluctuation of the component tangential to the actual wind speed is greater:

1 jet stream, polar and subtropic core cooperate, collective 8
2 polar jet stream , anticyclonical shearing (on the right of the core), collective 3

In some areas there seems to be an unisotropic turbulence. But the higher mean wind fluctuations found in this paper should not be interpreted at all events as "unisotropic turbulence" , because differences might be greater by flying airway across the jet stream core. When flying across the jet stream which is separated in a polar jet core and a subtropical jet core, higher mean wind fluctuations of component normal to the actual winddirection have been observed in the curvature of the current. If one crosses a more zonal frontal zone, higher mean wind fluctuations may be observed in the component tangential of the stream lines. These effects are given by a cooperative jet stream (collective 8).

The high mean wind fluctuation of the anticyclonic shearing (see figure 7, page 74, collective 3) is declared in the following. The effect of an air particle disturbed in geostrophic equilibrium is greater in the anticyclonic shearing than in the cyclonic shearing. Convective inital momentums arise by high energy of lability. In upper layers this energy of lability is strong, if cool air-masses of polar region flow over the warm water of the golf stream. The situation happens more often near the anticyclonical shearing of the polar jet stream than near the anticyclonical shearing of the subtropic jet stream.

The standard deviation of the vector differences (actual flight winds - geostrophic winds) is moderate in the marginal areas of the jet streams, especially in the anticyclonic shearings. High standard deviations of the vector differences (actual flight winds - geostrophic winds) are found in areas between two jet stream cores. These high differences depend on errors in analysing upper-level charts and the high "ageostrophic winds" in the cores. In most areas over the Atlantic the position of the jet stream cores is difficult to analyse.

3.6.4 STATISTICAL ANALYSIS OF THE FLUCTUATIONS AND AGEOSTROPHIC COMPONENTS OF THE UPPER WINDFIELD IN THE "PARALLEL SCALE"

The statistical checking of the data in a s-formed stream model called "parallel scale" shows no higher mean wind-fluctuation in troughs. In some instances high wind differences may be demonstrated in troughs, too (see chapter 5, Weber (70)), but there must be expected a higher meso-scale and micro-scale changing of the windfield at anticyclonic curvature. Important wind fluctuations have been observed at anticyclonic curvatures in areas with small radius of the stream lines (figure 8, page 74).

The differences between the actual flight winds and the geostrophic winds in areas with anticyclonic flow are a little higher than in areas with cyclonic flow. The discussion of cyclostrophic winds has shown more satisfactory results near the end of the anticyclonic current. At the beginning of the anticyclonic current the standard deviation of the vector difference between actual flight wind and cyclostrophic wind is much stronger than those of the vector difference between actual flight wind and geostrophic wind. Therefore the cyclostrophic windfield is especially uncertain at the beginning of the anticyclonal current. There may exist a systematical acceleration of an upslide motion in these regions.

Abstract Chapter 4 (page 79 - 88)

THE ACCURACY OF THE UPPER WIND FIELD DERIVED FROM THE 300 MB AND 200 MB CHARTS

During a flight (Montreal - San Francisco) it was possible to correlate the recorded data by Boeing 707-430 D-ABOD with the data measured by radiosonde of the aerological station Flint (72.637). The vector differences between radiosonde wind and flight wind (neighbouring data in computed grid system, sequence of dots 6 minutes) both measured in 200 mb amounted to 1.7 m sec^{-1}. Averaging flight winds over a distance about 100 km the vector difference decreased to 0.5 m sec^{-1}.

The pressure field can be analysed in the upper level charts only with uncertainty for the distance of the aerological station is more than 100 km. Therefore the data taken from upper level weather chart show no meso structure or micro structure of the atmosphere. The differences between geostrophic and flight winds might be reduced when a closer network of radiosonde stations is applied. In this case the weather charts might reproduce qualities of the meso scale.

The standard deviation of the vector differences between flight wind and geostrophic wind is composed by:

1 Errors in measurement by radiosondes
2 Errors in transmission and notation
3 Errors in analysing
4 Inaccuracy caused by large distances of the aerological network and fluctuations of the wind
5 Inaccuracy caused by movements of pressure fields, jet stream, isallobars, isobars, wind field etc., time lag between the measurements by aircraft and by radiosonde
6 Heterogeneity of measuring method (horizontal integration of wind measurement with aircraft, vertical integration of wind measurement with radiosonde)
7 Errors in measurement by aircraft
8 Adataption, ageostrophic winds, meteorological processes

Berggren (4) has described point 1 - 3. Point 4 is taxed by the value of the mean wind fluctuation. Point 5 may be calculated by arranging the flight data to the time lag of aerological measurement. Point 6 is less than 1 m sec^{-1}, found by correlating flight winds and radiosonde winds. Point 7 is discussed in chapter 9. The attained accuracy is less than 2.9 m sec^{-1}.

The geostrophic and flight wind data were smoothed by equations (3.3.1) - (3.3.3) described in chapter 3.3 and averaged by equation (3.3.4) (page 53). The results computed by these equations show a minimum in standard deviation of the vector differences between "smoothed flight wind and geostrophic wind" about a distance of 600 km (48 minutes of flight time) in accordance with the distance of Atlantic radiosonde stations (data, see page 83, Table 2: Correlation factor; page 86, Table 4: Standard deviation of the vector differences between geostrophic wind and flight wind).

The correlation factors of the vector differences between flight winds and geostrophic winds vary between 0.64 and 0.80. The correlation factors of the north component are more favourable than those of the east component. The correlation factor of the differences between flight wind and geostrophic wind is higher than the correlation factor of the differences between flight wind and cyclostrophic wind. The inaccuracy of the cyclostrophic wind is increased by the curvature of the contours. The maximum of the correlation factor has been computed for data smoothed over a distance about 600 km (see page 83, Table 2).

The standard deviation calculated from squares of the vector differences between flight wind and geostrophic wind is 15.5 m sec^{-1}, those between flight wind and cyclostrophic wind is 17.0 m sec^{-1}. The standard deviation is more than 14.0 m sec^{-1} if the time difference between flight wind and radiosonde wind is not less than one hour. For time differences less than one hour the standard deviation is 10.9 m sec^{-1}. The data of standard deviation shown in Table 3 must be reduced about 1 m sec^{-1} (mean instrumental error less 3 m sec^{-1}) in order to receive the mean error of geostrophic wind calculated from weather charts. This mean error of single measurement is composed mainly by meso- and micro structure and inaccuracy by analysing the areas of high wind speed and in second hand by errors in macro scale analysis and instrumental errors of the radiosonde.

The standard deviation of the vector differences between flight winds and geostrophic winds is about 9 m sec^{-1} for the collective containing only south winds (see figure 3, page 59). Considering the propagation of errors the following values may be subtracted (least squares method):

mean error, arising from large distance of aerological network	1 - 2 m sec^{-1}
mean error, arising from time lag between measurement by aircraft and by radiosonde	2 - 3 m sec^{-1}
mean instrumental error of flight wind	3 m sec^{-1}

Considering this estimation of error, the mean error of the geostrophic wind (for single data in upper level weather charts) is about 8 m sec^{-1}. The meso scale wind fluctuation is more than 4 m sec^{-1}. Therefore the mean standard deviation of the vector differences between flight winds and geostrophic winds cannot be less than this value. By using flight winds, the mean error of a single geostrophic wind (from weather chart) is in the limits of error between 4 m sec^{-1} and 8 m sec^{-1}, if the distance of the wind data along the flight path is less than 100 km. These advancements may be realized by an integrated electronic data processing system which encloses automatic recall, data collection, transfer, reduction, and international interchange. The accuracy of upper level charts can be increased by a compact network of observations. Nowadays the mean error of a geostrophic wind is more than 10 m sec^{-1} (single value, taken from upper level charts). The mean error increases to 15 m sec^{-1}, if the time lag between flight and radiosonde data is more than one hour. The mean error of the geostrophic wind increases near the tropopause and in areas of the jet stream.

These strong mean errors and standard deviations are important for the meso structure of the atmosphere and for single measurements. It is not established that geostrophic winds which are calculated as a mean value of an area, have the same accuracy in macro scale as in meso scale. In most cases the macro structure of the atmosphere is analysed correctly, but the location of small low pressure centres and the exact position of the jet stream is analysed with uncertainty. The distance of the Northatlantic aerological stations is large enough to modify the analysed charts and to adjust the flight winds exactly. A small graphic displacement of the jet stream core or a changing of curvature are sufficient to make the geostrophic or cyclostrophic winds similare to the flight winds.

Abstract Chapter 5 (page 89 - 100)

DISCONTINUITIES IN THE UPPER WIND FIELD

Some meteorological institutes describe atmospheric planes of shear by lines of isopotential with a break. These shear trajectories are not stationary but moving with the velocity of the wind component normal to the layer. The radius of curvature may be a few kilometres in the trough line at low wind speed. Actually the flow lines are not broken but strongly bent. In upper level charts of small scale planes of shear may be marked by lines of curvature with a break.

When an aircraft is crossing a plane of shear with strong differences of wind or a discontinuity of flow, the lifting forces change irregularly. The vertical acceleration of an aircraft is less by horizontal gusts than by vertical gusts but the effects of horizontal differences in the wind field should be scrutinized especially with regard to the supersonic air-traffic.

The meso structure of atmosphere is important for aero-space projects. The distinction of the atmospheric effects in macro-, meso- and micro-scale is the result of technical and not meteorological reasoning. Therefore the definitions (macro, meso, micro) may be changed in the future. The atmospheric classification defined by flight operational practices and flying qualities of propeller aircraft and jet aircraft, perhaps will not be valid for supersonic air traffic using aircrafts flying with 3 Mach. Nowadays the so called "meso structure" of the atmosphere measured by some subsonic aircrafts generates effects for supersonic traffic in the way as "clear air turbulence" acts in subsonic air traffic. The described wind fluctuations can produce unsteady flight airways for supersonic air traffic. These ways are not identified with such observed by subsonic air traffic. Disagreeable effects are produced to the supersonic air-traffic by discontinuities in the wind field. In these cases not only the lift but also the temperature of the pheriphery of the aircraft is changing.

Table 5 (page 94) shows the frequencies of the wind differences recorded in the periods of measurement in 1962 and 1963. The vector differences are computed for 6 minutes of flight time (about 100 km).

The collective has a Gaussian distribution. Strong wind shears or wind fluctuations were combined with clear air turbulence (unsteady of flight, vertical acceleration). But strong variation of the vertical acceleration is not always combined with strong wind fluctuations or changes of the meso structure of the wind field.

The mean wind fluctuation is $6\,m\,sec^{-1}$. Gradients of wind more than $18\,m\,sec^{-1}$ (three-fold mean wind fluctuation) over a distance about 100 km had been recorded in 1962 at 3 flights and in 1963 at 18 flights. Figure 9 and 10 (page 96 and 97) show the examples of 1963 in 12 upper level charts (300 mb). The centre of a chart shows the region with high strong wind differences. The data are given in Table 6 (page 98). The strong wind differences and fluctuations are measured in the relatively shell boundary zone of the jet stream and in the surroundings of a cyclone with small inclination of the axis. These upper level cyclones have a surrounding with low wind speed. The jet stream flows near by with high speed. In some cases the finger-shaped jet stream extends into the area with low wind speed. The atmosphere may produce a free jet but the discussed measurement cannot prove its evidence. Effects in low levels can produce strong differences in wind speed. Probably the aircraft will measure strong wind shears of the jet stream and strong initial forces of perturbations. The influences of low level effects on the upper wind field are more important than usually accepted.

Abstract Chapter 6 (page 101 - 107)

DAILY VARIATIONS OF THE UPPER WIND FIELD

A diurnal oscillation of the upper wind has been proved for different stations. The authors describe deviations from the mean value. However, the differences between geostrophic winds and those measured by aircraft (in the following called flight winds) during different hours of the day indicate a systematic oscillation. Figure 11 (page 103) shows the mean vector differences between geostrophic and flight winds arranged in view of GMT. The time is indicated at the end of each vector. The vectors have their source in the origin of the system of coordinates. They must be added to the geostrophic wind in order to get the flight wind. One vector represents the mean of the differences measured between the indicated time and the following one.

Figure 12 (page 105) shows the mean vector differences between geostrophic and flight winds arranged in view of local time. Figure 13 (page 105) shows the same after eliminating of vector differences greater than 20 m sec^{-1}. The ageostrophic vectors change only a little. The figures indicate a diurnal oscillation of the ageostrophic wind. Between 9 and 19 hrs local time an additional southern component must be added to the geostrophic wind (taken from the weather chart) whereas between 20 and 7 hrs an additional northern component must be added to get the mean flight wind. One has to add a northern component of 1 m sec^{-1} to the daily mean of the geostrophic wind to get the daily mean of the flight wind. The sum of the zonal components is zero.

If the data are arranged in view of local time the Fourier analysis for the northern component of the vector differences produces only an oscillation with a period of 24 hours having an amplitude of 3.9 m sec^{-1}. The maximum has been stated at 3 hrs. The minor periods found by other authors do not exceed the noise level of 1 m sec^{-1} concerning the ageostrophic winds in the northern component. As to the eastern component the 1st, 2nd and 4th harmonics of the diurnal oscillation have the following amplitudes:

2.8 m sec^{-1} for the 1st harmonic
3.6 m sec^{-1} for the 2nd harmonic
2.4 m sec^{-1} for the 4th harmonic.

The 1st harmonic of the northern component of the ageostrophic wind is caused by the diurnal oscillation of the temperature and the transfer between the tropical and polar regions. The southern components of the ageostrophic wind found for the day hours cause an increased momentum in the westwinds whereas during the night the northern component of the ageostrophic wind cause a decrease of momentum. In the stratosphere there are indications of a stronger ageostrophic wind towards the pole, during the day hours, than in the troposphere. Such winds cause subsidence by the poleward shrinking of the atmosphere. Right above the tropopause the maximum southern and also western component of the flight wind can be stated in the mean value. Strong inversions are generated by gliding-off processes in upper layers (Raethjen (41)). The tropopause is the strongest temperature inversion in the atmosphere and has a similar origin.

Abstract Chapter 7 (page 108 - 124)

DISCUSSION OF SOME ESPECIALLY FLIGHTS

7.1 WEATHER SITUATION AND RECORDED DATA OF THE 6.4.1962

Figure 14 (page 108) shows the 200-mb-upper-level-chart on 6.4.1962 at 00 GMT. The line printed in bold indicates the flight of Deutsche Lufthansa Boeing 707-430 D-ABOD from New York to Frankfurt. The tropopause had been crossed twice on an isobaric track. The first part of the flight (New York to 57° West) happened in the troposphere, after climbing the part from 57° West to 35° West happened in the stratosphere, the part from 35° West to Ireland in the troposphere, and the last part in the stratosphere again. The first isobaric crossing of the tropopause happened on the subtropic side of the jet stream, the second on the left side of the high wind speed area on the polar side of the jet stream.

Figure 15 (page 110) shows the data recorded by aircraft and measured by radiosondes. The line printed in bold indicates the tropopause (dotted line tropopause analysed by aircraft measurements and isentropics). The flight level is drawn as a thin straight line. The roughly straight lines (parallel to the abscissa) show the sectional planes of isentropics. The thin straight lines normal to the abscissa mark the ascents of radiosondes. In the part above the recorded flight data are shown. The course of curve represents the temperature recorded by aircraft. The parts of the flight with bumpiness (CAT) recorded with aircraft are marked in the diagram by bold printed beams. The short wind arrows indicate the flight wind in a distance of 6 minutes of flight. The wind arrows (bold printed) show mean winds computed from the records integrated over a distance of about 20 minutes of flight. The diagram shows that the data of temperature and wind measured by aircraft and radiosonde are conform.

Figure 16 (page 111) shows a vector diagram of the vector differences between mean wind data (integrated about 20 minutes) and actual flight winds (measured directly, not smoothed). In figure 16 the straight line from origin to a mark (+ troposphere, x stratosphere) indicates the absolute value of modulus of the vector difference. The variation of the wind (in this case given by vector difference between 20-minutes-mean and single data) are preponderant less than 10 m sec^{-1} and only in some cases exceed more than 20 m sec^{-1}. For analysing upper level charts the mean flight winds are more adapted than single actual data (not smoothed).

Figure 17 (page 112) gives a diagram of the cross-flight wind components (above) and the recorded temperature of the flight New York - Europe on 6.4.1962. The abscissa shows the geographical longitude, the ordinate gives the cross flight wind component. The scale on the left side is valid for the cross wind components, the scale on the right side of the diagram is valid for the temperature recorded with aircraft.

Thin solid line: cross wind component computed out of Doppler data, measurement by aircraft (Q_f)

Thin dashed line: cross wind component of the flight geostrophic wind computed out of measurements with aircraft (Radio Altimetry) (Q_r)

Thin dotted line: cross wind component of the geostrophic wind computed out of upper level chart (Q_k)

Thick solid line: difference between mean flight cross wind (Q_f) and flight geostrophic cross wind (Q_r) (both measured by aircraft)

Thick dotted line: recorded temperature (below)

At the flight on 6.4.1962 on the polar side of the subtropic jet stream the measured flight winds had been smaller than geostrophic winds in cyclonal curvature. On the side of anticyclonal shearing (south of jet stream core) the measured flight winds had been greater than geostrophic winds in anticyclonal curvature shown in figure 17. The jet stream had been moved to the anticyclonic shearing in those days.

Crossing the tropopause at 35° West the wind speed maximum is combined with low temperature (before temporary rise) and low ageostrophic cross wind component (difference $Q_f - Q_r$). The strongest ageostrophic components of the cross wind had been found in the cooler part of the troposphere traversed later in figure 17. In the beginning of the tropospheric part of the flight, the maximum of the geostrophic cross wind was recorded later than the maximum of the Doppler cross wind (see fig. 17; 6.4.1962). The maximum of the wind speed is situated on the polar side of the jet stream core, but the strongest gradient of pressure on the subtropical side of the jet stream core traversed later on in the flight on 6.4.1962.

Figure 18 (page 114) shows the record of temperature and flight wind (absolute value of modulus of vector) on 6.4.1962 with isobaric crossing of the tropopause. The abscissa sign the time (GMT). The ordinate gives the absolute value of wind vector in knots and temperature in $^{\circ}$C (low temperature above, high temperature below). The direction of wind is shown in the lower part of figure 18. At the crossing of the tropopause from stratosphere to troposphere the records show a sharp droop curve, short rise and again decreasing of temperature combined with a sudden shift of wind. The wind had been variing more in the last part of the stratospheric flight. Before crossing the tropopause the wind velocity decreased and ascended suddenly during the crossing up to a maximum of 42 m sec^{-1} (combined with lowest temperature). In the upper troposphere strong wind fluctuations are combined with bumpiness (Clear Air Turbulence). After 300 km flight distance in the troposphere the fluctuations of wind and CAT had been decreased.

At 35° West crossing the tropopause isobaricly the variation of temperature was about 10°C, but near England crossing the tropopause again no strong differences in temperature had been recorded. The high differences of temperature at the tropopause on the west side of the anticyclonal field of flow may be favoured by emission at night and by downslide motions of stratospheric airmasses above the tropopause. Figure 19 (page 115) shows in a vertical-plane on the meridian of 35° West a slope of the isentropics to the north (above ocean vessel E,D,C,A, meteorological station 04.360). On the easterly side of the subtropic ridge of high pressure no slope of the isentropics is remarkable on the position of the aircraft in the stratosphere. The slope of these isentropics is situated more in the south. The current in this region had a stronger north component. Figure 20 shows the vertical-plane on the meridian of 5° West.

Crossing the tropopause isobaricly at 35° West the records of temperature have shown a local lowering of the isentropics on the tropospheric side and a local lift of the isentropics on the stratospheric side. The veering of the wind started with onset of decreasing low temperatures. A sudden backing into the preceding wind direction is remarkable on the lowest position of the local low of the isentropics.

7.2 WEATHER SITUATION AND RECORDED DATA OF THE 26.3.1963 - 28.3.1963

Figure 21 (page 117) shows the weather chart of 200 mb and figure 22 (page 118) shows the weather chart of the surface of 27.3.1963-00 GMT. The recorded data are put along the flight via Frankfurt, London, South of Greenland, Labrador, Montreal, Denver to San Francisco. Over the Atlantic the first part of the flight happened in the stratosphere, later on over the American continent the aircraft has flown near the tropopause.

The recorded data have shown a decrease of temperature when the aircraft approached the tropopause from stratosphere to troposphere (27.3.1963 decreasing from -54.0°C to -59.0°C). Therefore one can locate an "isentropic hump" on the stratospheric side of the crossing of the tropopause. Within two minutes the temperature rose to the preceding level (27.3.1963 to -54.0°C). One can locate an "isentropic trench" on that place. Flying into the lower troposphere the temperature decreased again (27.3.1963 to -61.5°C). It may be possible that after the first crossing of the tropopause (low temperature) the aircraft had flown in the stratosphere (area of high temperature) again and after the second crossing the aircraft had flown in the troposphere. But the recorded data of wind and vertical acceleration suggested an interpretation as given above (see chapter 8, too).

CAT had been recorded suddenly and strongly flying into the "cool-case". After a short break oscillations in vertical acceleration had become stronger on the flight across the "isentropic trench". Crossing the tropopause the wind velocity had been strongest in the so called "isentropic hump" (on the stratospheric side, connected with the lowest temperature). The wind had been backing in the troposphere. The sections of CAT had been strong in relation with an isentropic layer.

Figure 23 (page 121) shows a photograph of the continuous cloud cover with little vortex formations over the easterly Northatlantic. The clouds dissolved over the cool water-masses of the East Greenland current (fig. 24, page 121) and of the labrador current (fig. 25, page 123). The stratocumulus clouds formed in cloudstreets near ocean vessel B. The cloudbanks had been oriented along the wind direction (from the Davis-strait around the Iceland cyclone to the Golfstream area). A wind maximum existed within the convective layer (about 700 mb) and an inversion on top of it. The spacing between these cloudstreets generally lies between 2 and 3 times the height. The convection had been generated by the warmer water in the region about ocean vessel B.

The circulation near the tropopause may be defined as a circulation in lower cloud streets. The analogy of circulation in stratocumulus, cumulus, cumulonimbus cloud streets of low layer, and of the band structure of altocumulus-clouds, cirrus-clouds in upper levels suggest to explain the circulation of the tropopause in the same way.

In the area South of Greenland the tropopause had a very low altitude and rapidly rose near Labrador. Figure 26 (page 123) shows the steepness of the tropopause and the rise of the cirrus on the right side of the photograph.

Abstract Chapter 8 (page 125 - 130)

CIRCULATION IN MICRO-SCALE AND MESO-SCALE NEAR THE TROPOPAUSE

Figures 27 to 30 (page 126) show a possible circulation near the tropopause (schematicly by using the flight data). The micro- and meso-circulation is indicated by arrows. The isobars are shown by dashed lines, the isentropics by thick solid lines. The layers of cool and warm air are differently hatched.

Figure 27 shows the temperature, recorded by an aircraft which is crossing the tropopause twice. The aircraft had flown on an isobar and parallel to the mean current, flying from the stratosphere into the troposphere and into the stratosphere again.

Figure 28 illustrates a horizontal profile of the atmosphere near the tropopause. The curve in bold-face type marks the cutting line of the tropopause. On the troposphere side of the tropopause the areas with cool air and warm air are differently shaded. The flight-track and the isobars are drawn straightly. The isobars are shown by dashed lines and indicated with P_0, P_1, P_2, P_3 (from low to high pressure).

Figure 29 shows a vertical profile of the atmosphere. The figure is orientated parallel to the mean current and to the flight-track. The figure characterized the crossing of the tropopause shown in figure 28 on left hand. The tropopause is marked in bold-face type. The areas different shaded illustrate the very cool level (skin of troposphere) and the lower warm level below of the tropopause. The straight line defines the flight level. The curves show the profile lines of the isentropics analysed by aircraft records and in comparision with the measurements of radiosondes.

Figure 30 gives a vertical profile of the atmosphere perpendicular to the mean geostrophic wind and to the track. The tropopause is drawn in bold-face type. The thin lines mark isentropic surfaces.

The following characteristics can be observed and measured, if an aircraft is crossing the tropopause on an isobaric layer.

1 Entering the troposphere from the stratosphere a small very cool area (about 20 km in the wind direction) is situated behind the warm air of the stratosphere. A warm area (about 20 - 30 km long) is following. Behind this phenomenon there is an area of low temperature in the upper troposphere. It is possible that the aircraft is crossing small horizontal layers of different temperatures

2 The maximum of flight wind is to be found in the low stratosphere, frequently in the small cool layer.

3 The flight wind is veering in the cool layer of the tropopause ("isentropic hump"), i.e. in the direction of the polar flank of the jet stream.

4 The flight wind is backing in the warm layer ("isentropic trench") which is lower than the cooler one with respect to the tropopause.

5 The strongest vertical acceleration (probably released by vertical gusts) is situated in the area of the strongest dropping of the isentropic surfaces (warm "isentropic trench").

6 The shocks are short and frequently isolated in the "isentropic hump" at the tropopause (probably released by horizontal gusts).

E.R. Reiter (51) had discussed that a warm area (so called "isentropic trench") is located on the cyclonic flank of the jet stream. In this area stratospheric air drops into the troposphere. South of the jet stream core E.R. Reiter had found a tongue of cold air which is explained by ascending air masses. The two phenomena, "isentropic trench" and "isentropic hump", had been related by Reiter to the position of the jet stream and its fingers. These appearances have a larger scale (macro scale) than the described one and extend the structure of the tropopause shown in figure 29 . The "isentropic trench" (so called by Reiter) is located on the polar stratospheric side of the tropopause (figure 29, on left hand). This phenomenon should better be called "isentropic trough". The macro scale "isentropic hump" (so called by Reiter) is located on right hand (figure 29) of the sketched meso scale "isentropic hump". The macro scale hump should be called "isentropic hill" .

Near the tropopause the radiation processes and effects of current and exchange influence the temperature of the layers beyond the tropopause. The tropopause may be considered as an inversion of downslide motion. This downslide motion is produced by stratosphere air masses flowing from the tropics to the polar region across the zone of west drift (Raethjen (41)). The tropopause is the "skin" of the troposphere (upper layer with higher humidity and dust) and emits into the outer space without an atmospheric back radiation. Therefore low temperatures are located at the tropopause. The adiabatic descending of cool air particles may generate a warmer lower level. Below the tropopause a maximum of wind speed is located in the thin cool upper layer. Frequently the current of the upper layer is veering, that of the lower layer backing.

Generally there is a decrease of the isentropic levels from the subtropics to the polar region. Therefore the cool veering current is upsliding rather than the lower warmer current (which frequently may have a downslide motion). In the lower level (located below of the warm layer with downslide motion) a new inversion may be generated by downslide motion (descend of the tropopause, multiple tropopause, laminated).

Near the tropopause isobaric levels are crossing isosteric levels. Figure 28 (horizontal plane) shows a possible structure of the field of pressure and density near the tropopause. The micro- and meso scale air currents produced by the fields of temperature and of density may not compensate the field of pressure. On the left hand (upward of the macro scale current) of figure 28 a cyclonal, on right hand (down of macro scale current) an anticyclonal meso circulation may be produced below the tropopause.

Figures 29 and 30 show vertical planes parallel and perpendicular to the mean (macro scale) current. The tropopause may be defined as layer but not as level regarding the meso scale circulations described in this paper. The tropopause encloses a thin cool upper layer and warm lower layer and the circulation between both layers. The tropopause is generated continually on the top of the cool layer (level of emission).

The following model of circulation could describe the records and observations made on flights across the tropopause (parallel to the geostrophic wind):

> The upper layer of the tropopause produces cool air by radiation. The cool air drops downstream (into the "macro scale winddirection"), is warmed up by adiabatic subsidence and flowing isentropically in the direction of the polar flank of the jet stream. This warmer air tends to ascend against the "macro scale wind" (fig. 28-30) if the upper layer becomes cooler (vertical lability).

This process produces helicoidal currents in micro- and meso scale (Prandtl (39)) with a band structure of cirrus-clouds and clear air turbulence. Kuettner (33) has studied cloudstreets in the jet stream. The band structure of clouds near low level inversions was described by Seilkopf (61).

The turbulent motions in clouds and at inversions may be explained by arranged micro scale systems of circulation. Considering the micro scale circulations near the tropopause in a macro scale they appear to be turbulent. Examining then in meso scale, they seem to be of an ordered meso structure of the circulations (observed by aircraft). The macro structure of the atmosphere is arranged by the systems of high and low pressure (observed by radiosonde). The knowledge of the scales (macro, meso, micro) of arranged processes is a function of the measuring system.

Abstract Chapter 9 (page 131 - 148)

METHOD AND ACCURACY OF WIND MEASUREMENTS BY HIGH SPEED AIRCRAFTS

The Boeing 707-430 aircraft D-ABOD of Deutsche Lufthansa was equipped with a Doppler-radar-system in conjunction with a highly accurate air-speed indicator for measuring the direction and speed of the horizontal component of the wind. The ground speed was measured by the Doppler shift of a radar beam reflected from the earth's surface. Drift produced by winds is also measured by means of Doppler effects.

To measure the air temperature during flights of high speed aircraft a Rosemount-Thermometer (Model 102 Series) was used. The probe measured the Total Air Temperature defined as the sum of the Static Air Temperature plus the total adiabatic rise of temperature of the air due to the effect of relative motion of the airplane through the air. The data of flight wind discussed in this paper are computed from Static Air Temperature, Machnumber and True Air Speed. The Static Air Temperature is defined as the air temperature measured by a stationary thermometer in static air, at the altitude of the aircraft.

The SFIM-recorder A 2610 gives a record on photographic paper film of the following meteorological and aeronautical parameters:

1 Total Air Temperature
2 Vertical Acceleration
3 Magnetic Heading
4 Drift Angle
5 Groundspeed
6 Indicated Airspeed
7 Pressure Altitude

The parameters had been checked. The Doppler-wind-data compared to each Rawin data, have sufficient accuracy. The accuracy and the limits of the record are given in Table 7 (page 133). The method of computing the accuracy of the wind data measured by means of Doppler radar during flights of high-speed aircraft is described in chapter 9.2 (page 136) and 9.3 (page 139). The high accuracy of Doppler-wind-data, computed for an equal damping of the recorded parameters, can be seen in Table 8 (page 134).

2. Probleme der Messung und Verarbeitung meteorologischer Daten aus Düsenverkehrsflugzeugen

Die Meteorologie versucht die Struktur und die physikalischen Vorgänge in der Atmosphäre zu erkennen und zu beschreiben. Die unteren Dekameter der Atmosphäre sind dem Meteorologen für Serienmessungen leicht zugänglich. Vergleichbare Serien können dagegen in großen Höhen schwer gewonnen werden. Geeignete Träger für die Instrumente sind der Ballon, vertikal aufsteigend oder auch horizontal driftend, das Flugzeug, die Rakete und der Satellit.

Die neuen Träger für Instrumente ermöglichen es, die Beobachtungen des bestehenden Radiosondennetzes wesentlich zu verdichten und auch auf die Südhalbkugel der Erde sowie insbesondere auf die Ozeane auszudehnen. Sie gewinnen ferner Beobachtungen aus höheren Schichten der Atmosphäre, können eventuelle kosmische Einflüsse besser feststellen und auch globale Messungen über die Strahlungsverhältnisse und den Wärmehaushalt der Erde durchführen. Die Fortschritte der Meteorologie hingen in der Vergangenheit wesentlich von der jeweiligen Erweiterung der Beobachtungen und der Vermehrung der Stationen ab. Bis vor etwa 100 Jahren schenkten die Forscher dem Windfeld gegenüber dem Druckfeld die größere Beachtung. Diese Ansichten wandelten sich mit der Vermehrung und besseren Übermittlung der Druckbeobachtungen. Als es möglich wurde, Isobarenkarten zu zeichnen, gab man dem Luftdruckfeld den Vorrang und empfand die Zyklone nicht mehr als "Wirbel" sondern als "barometrische Depression".

Eine ähnliche Entwicklung wiederholte sich bei der Erfassung der dritten Dimension. Das Windfeld der höheren Schichten wird als "geostrophisch" angesehen und das Druckfeld so angelegt, daß die Isohypsen nach den Druckwerten der aerologischen Stationen parallel zu den beobachteten Windvektoren gezeichnet werden. Diese Konzeption bürgerte sich durch den

Mangel an Beobachtungen aus höheren Schichten ein, obwohl bekannt war, daß jede Veränderung des Druckfeldes nur durch ageostrophische Komponenten des Windfeldes erzeugt werden kann.

Neue Impulse für den Fortschritt der meteorologischen Forschung sind durch die neuen Träger für Instrumente, wie das Flugzeug, die Rakete und den Satellit zu erwarten. Es verdient das Windfeld zunächst die größere Beachtung, weil die Beobachtungen und Messungen des Windes sehr viel besser verdichtet werden können.

Satelliten liefern den Wetterdiensten Aufzeichnungen der Wolkenfelder und Messungen der Strahlungsverhältnisse. Die Zeit ist noch nicht reif, aus diesen Beobachtungen auf die Windverhältnisse in verschiedenen Höhen zu schließen.

Raketen dienen als Träger für Instrumente in große Höhen. Meßergebnisse durch Raketen aus der Troposphäre und Stratosphäre sind kaum in Serien vorhanden. Eigenarten des Windfeldes können an dem Verhalten der Rauchfahne nach dem Start studiert werden.

Driftballone werden auf der Südhalbkugel routinemäßig im Wetterdienst zur Analyse des Strömungsbildes verwandt. Sie haben gegenüber Flugzeugen den Nachteil, daß ihr Flug schwer verfolgt und nicht gesteuert werden kann.

Flugzeuge verkehren zahlreich zwischen den Druckniveaus 200 mb und 300 mb. Über den Ozeanen wird heute kaum noch in den niederen Schichten geflogen. Die Flugzeuge ergänzen als Meßträger die Radiosonden, welche viermal täglich vertikale Schnitte durch die Atmosphäre liefern. Die Übereinstimmung der nach den beiden Verfahren ermittelten Werte muß nachgewiesen werden.

Die großräumige Zirkulation wird gut durch das bestehende Radiosonden-Netz erfaßt. Die Messungen mit dem Flugzeug liefern bessere Einblicke in die kleinräumigen Strukturen der Atmosphäre, weil sie isobar kontinuierlich gemessen werden können. Die Beobachtungen und Messungen aus Flugzeugen sollten sinnvoll den Wetterdiensten zugänglich gemacht werden.

Die meteorologischen Dienste haben schon sehr früh die Luftverkehrsgesellschaften zur Beobachtung und Abgabe bestimmter meteorologischer Elemente angehalten. Leider werden die Möglichkeiten, die sich aus einer gut organisierten Beobachtung und Registrierung der Wetterbedingungen längs eines Flugweges ergeben, nicht genutzt. Hindernisse liegen sowohl bei der Messung und Registrierung im Flugzeug als auch bei der Übermittlung an Bodenstationen und der anschließenden Datenverarbeitung im internationalen Austausch.

Die vielseitige Verwendung der erflogenen Daten wird eingeschränkt, weil alle Flugzeuge zwischen zwei Kontinenten längs des Weges minimaler Flugzeit fliegen. Start- und Landezeiten müssen den Wünschen des Fluggastes entsprechen. Daraus ergibt sich eine Bündelung der planmäßigen Verkehrsmaschinen zu bestimmten Zeiten in gleichen Gebieten. Die Wetterdienste suchen dagegen Daten, die gleichzeitig beobachtet und über ein möglichst großes Gebiet verteilt sind. Da die Flugzeuge über lange Strecken isobar aber nicht in Hauptisobarenflächen fliegen, ergeben sich weitere Schwierigkeiten für die vertikale Zuordnung der Beobachtung oder Messung in eine analysierte Fläche.

Die vorliegende Arbeit zeigt, daß über Registriergeräte brauchbare flugmeteorologische Daten zu gewinnen sind. Auch dem täglichen Beratungsdienst würden wertvolle Informationen über isobare Schnitte durch die Atmosphäre geliefert, sofern während des Fluges die Geräte bodenseitig abgefragt werden könnten.

Die heute eingesetzten elektronischen Rechenmaschinen würden sowohl im Speichervermögen als auch in der Rechengeschwindigkeit ausreichen, die vom Flugzeug abgefragten Werte zu speichern und mit anderen Daten zu vergleichen. Sie könnten darüber hinaus für eine vorgegebene Zeit die Wetterlage berechnen. Die notwendigen elektronischen Speicher für die Flugzeuge sind in Satelliten erprobt. Die zunehmende Verkehrsdichte, die technische und navigatorische Überwachung des Flugweges und die Flugkontrolle zwingen von bisherigen Methoden abzugehen und moderne Datenverarbeitungsverfahren einzuführen.

Neben die erwähnten Schwierigkeiten, wie Bündelung der Flugrouten, gleiche Flugzeiten und die erforderlichen Geldaufwendungen für die Datenverarbeitung, treten Erschwernisse meßtechnischer Art am Flugzeug. Die in der Grenzschicht zwischen Flugkörper und umgebender Luft gemessenen Werte sind nicht notwendig charakteristisch für die ungestörte Atmosphäre. Ebenso muß nachgewiesen werden, inwiefern die heute benutzten Fluginstrumente und Meßfühler makro-, meso- oder mikrostrukturelle Erscheinungen der Atmosphäre brauchbar für den täglichen meteorologischen Beratungsdienst feststellen können.

Wir folgen in der vorliegenden Arbeit ungefähr der strukturellen Einteilung der Atmosphäre, wie sie Reiter in "Meteorologie der Strahlströme" (50) beschreibt. Die Makrostruktur der Atmosphäre (Größenordnung über 300 km horizontaler Erstrekkung) wird wesentlich durch das vorhandene Radiosonden-Netz erfaßt. Die mit Flugzeugen "erflogenen" Winde ergänzen diesen Bereich. Feinere Aufschlüsse lassen diese Messungen im mesostrukturellen Bereich (zwischen 10 km und 300 km horizontaler Erstreckung) erwarten. Die Mikrostruktur der Atmosphäre (unter 10 km horizontaler Erstreckung) äußert sich für den Flugverkehr in oft unangenehmen Erschütterungen oder kurzperiodischen Schwankungen des Flugkörpers. Diese Erscheinungen werden in der Luftfahrt unter dem Namen "Clear Air Turbulence" (CAT) zusammengefaßt.

Mit dieser Auslegung des Wortes "Turbulenz" hat man sich von der hydrodynamischen Konzeption der Turbulenz in der nicht-laminaren Strömung entfernt.

Die Stärke der "CAT" wird am Grad der Erschütterungen des Flugkörpers gemessen. Dadurch wird "Clear Air Turbulence" abhängig von der Konstruktion des Flugkörpers und seinen Flugeigenschaften in gestörter Atmosphäre. Verschiedene Flugzeugtypen reagieren bei gleicher Böigkeit unterschiedlich. Die Heftigkeit der Erschütterungen im Flugzeug wird von der Eigengeschwindigkeit, dem Anstellwinkel und der Beladung des Flugzeuges (Flächenbelastung) bestimmt. Die Beratungsdienste werden dieses bei Einführung des Überschallflugverkehrs beachten müssen, weil dann in gleichen Niveaus Düsenverkehrsflugzeuge herkömmlicher Bauart und Überschallflugzeuge während der Unterschallphase ihres Fluges nebeneinander über längere Strecken verkehren werden.

Zur Sammlung von Erfahrungen und zur Untersuchung der meteorologischen Ursachen der "CAT" in der Operationshöhe des Düsenluftverkehrs wurde in eine Verkehrsmaschine der Deutschen Lufthansa vom Typ Boeing 707-430 mit dem Kennzeichen D-ABOD ein Vielfachschreibgerät der Firma SFIM eingebaut. Der folgende Text schildert die damit gewonnenen Aufschlüsse und Erkenntnisse über die Mesostruktur des Windfeldes. Das Forschungsprojekt versuchte nicht, Daten zu bestimmten ausgewählten Wetterlagen oder Typen des jetstream zu gewinnen. Von März 1962 bis Mai 1962 und von März 1963 bis Oktober 1963 konnten auf 143 Atlantik-Überquerungen 8968 einwandfreie Windmessungen in einem zeitlichen Abstand von sechs Minuten erflogen werden. Die Daten des Jahres 1962 wurden statistisch mit geostrophischen Winden aus der Wetterkarte verglichen.

3. Horizontale Fluktuationen des Höhenwindfeldes

3.1 Vorbemerkungen

Die folgenden Kapitel beschreiben die Struktur der Atmosphäre. Dabei handelt es sich um kleinräumige Veränderungen der atmosphärischen Zirkulation, die von keiner der gebräuchlichen Wetterkarten erfaßt werden. Die Luftteilchen erleiden gerade in mikro- und mesostrukturellen Bereichen starke Beschleunigungen. Es ist zu erwarten, daß dort ageostrophische Komponenten auftreten. Heute bleiben die mikro- und mesostrukturellen Erscheinungen der Atmosphäre für die Analyse und Vorhersage der meteorologischen Felder unbeachtet und werden als "meteorologischer Lärm" in numerischen Modellen ausgesondert. Der in den nächsten Jahren zu erwartende Überschallflugverkehr verlangt die genaue Kenntnis und Vorhersage der atmosphärischen Mesostruktur.

Die meso- und mikrostrukturellen Vorgänge in der Atmosphäre können durch verfeinerte barokline Modelle beschrieben werden. Heute liefert kein Rechenmodell, welches wesentliche barokline Eigenschaften der Atmosphäre berücksichtigt, sehr viel bessere Ergebnisse für die Praxis als barotrope Modelle. Auch die Benutzung verschiedener Filtermethoden führte bisher zu keinen wirklichen Verbesserungen. Der Grund dieses Übels ist die Atmosphäre, deren Makrostruktur wesentlich barotrop, deren Mikro- und Mesostruktur dagegen baroklin ist.

Obwohl die Modelle mit baroklinen Verbesserungen den barotropen Modellen nicht wesentlich überlegen sind, darf die Bedeutung der Thermodynamik, speziell der Konvektion, für die allgemeine Zirkulation und deren "Störungen" nicht unterschätzt werden. Dem Synoptiker bleiben thermodynamische Vorgänge wegen der Maschengröße des aerologischen Netzes von ungefähr 300 km, über dem Nordatlantik sogar mehr als 1000 km, und des zeitlichen Abstandes der Beobachtungen von 12 Stunden verborgen, er spürt die Wirkungen in Fehlurteilen über die

Wetterentwicklung. In einem rotierenden Wassermodell kann die Thermodynamik der Atmosphäre durch Mechanik nachgebildet werden, indem Pumpen die Vertikalbewegungen erzeugen. Ein Massenteilchen, dessen geostrophisches Gleichgewicht in diesem Modell durch einen Initialimpuls gestört wird, vollführt eine horizontale Störbewegung, die sich aus dem Durchlauf der Trägheitsellipse und der geostrophischen Bewegung des Systems zusammensetzt. Da in dem Wassermodell die Quelle für Initialimpulse ortsfest gehalten werden kann, verformt sich die Strömung bei Einhaltung bestimmter Bedingungen hinter der Quelle antizyklonal.

Die Atmosphäre bietet nicht immer ortsfeste stark wirkende Quellen für "Initialimpulse", die in der Makrostruktur der Strömung nachgewiesen werden können (Beispiele hierzu siehe Raethjen (43), Höflich (27)). Ein gleichförmiges geostrophisches Feld wird von jedem hinreichend starken vertikalen Impuls merklich gestört; sofern die aufsteigenden Luftteilchen mit einer anderen horizontalen Geschwindigkeit in das betrachtete Niveau gelangen. Konvektive Initialstörungen sind in der Atmosphäre mannigfach vorhanden. Jeder sich entwickelnde Cumulus muß das höhere Niveau stören. Genauso stören strahlungsbedingte vertikale Austauschvorgänge an der Tropopause und Austauschvorgänge durch Windscherungen das horizontale geostrophische Stromfeld. Es kann gar nicht ein gleichförmiges ungestörtes Stromfeld in höheren Schichten existieren, sondern es muß in höheren Schichten mit starken örtlichen und zeitlichen Schwankungen des Feldes gerechnet werden.

Es ist gefährlich, alle nicht durch Radiosonden bestätigte und nicht in der Wetterkarte erkennbare Vorgänge als unbedeutend zu bezeichnen. Die Größe der Reynoldschen Zahl (Verhältnis Beschleunigungsarbeit zu Reibungsarbeit) läßt bei den Dimensionen der Atmosphäre überhaupt keine laminare Strömung mit höheren Geschwindigkeiten zu. Hier mag eine Schwierigkeit für alle Modelle liegen, die sich zunächst mit der

Errechnung beziehungsweise Herstellung eines quasistationären Grundzustandes begnügen, dem aber Störungskomponenten, meso- und mikrostruktureller Art überlagert sind.

Eigenschaften des Windfeldes, die außerhalb der Fehlergrenzen des Meßsystems liegen, können durch statistische Methoden nachgewiesen werden. Es muß dabei stets beachtet werden, daß die Statistik zwar mathematisch einwandfrei, das heißt überzufällig, eine Abhängigkeit zweier Größen darlegt, aber ein physikalischer Zusammenhang nicht notwendig bestehen muß. Diese Eigenschaft der mathematischen Statistik birgt gerade für Erklärungsversuche der komplizierten Vorgänge in der Atmosphäre viele Gefahren in sich. Es ist deshalb nützlich, zunächst einmal physikalisch überschaubare Vorstellungen zu entwickeln und erst dann mit Hilfe der mathematischen Statistik zu prüfen, ob die entwickelten Vorstellungen auch für die Atmosphäre zutreffen.

Die Erfassung der Mikrostruktur der Atmosphäre ist besonders schwierig, weil alle Meßträger in ihrer Flugbahn so stark gestört werden, daß die Anzeige meteorologischer und flugbetrieblicher Parameter nicht einwandfrei erfolgt; außerdem bereitet die notwendige zeitliche Auflösung Schwierigkeiten. Es ist deshalb naheliegend, die mesostrukturellen Vorgänge der Atmosphäre zu betrachten, um auf diese Weise zu einem besseren Verständnis der makrostrukturellen Vorgänge, der Beeinflussung der Makrostruktur durch kleinräumige Vorgänge und der Mikrostruktur zu gelangen.

In der vorliegenden Arbeit unterscheiden die Ausdrücke "mittlerer Fehler", "mittlere Streuung" und "systematischer Fehler" folgende in der Statistik gebräuchliche Ausdrücke:

1. Der Ausdruck "mittlerer Fehler" wird benutzt, wenn der Wert nach dem Fehlerfortpflanzungsgesetz von Gauss abgeleitet ist. Der mittlere Fehler des erflogenen Windes bedeutet einen nach dem Fehlerfortpflanzungsgesetz aus den mittleren instrumentellen Fehlern berechneten Wert.

2. Der Ausdruck "mittlere Streuung" steht für die Wurzel des Mittels, welches aus den Quadraten der gemessenen Abweichungen der Einzelwerte vom tatsächlichen (wahren), hier arithmetischen Mittelwert des Kollektivs gebildet wird.

3. "Systematische Fehler" sind bekannte Abweichungen des gemessenen Wertes vom tatsächlichen (wahren) Wert und werden als Verbesserungen zu dem Meßwert addiert.

Es wird folgende Definition eingeführt:

4. Für die Wurzel des Mittels, das aus den Quadraten der gemessenen Wind-Differenz-Vektoren zwischen zwei Winden an benachbarten Meßpunkten, die im Abstand von sechs Minuten Flugzeit (ungefähr 80 bis 100 km Entfernung) zu einander liegen, steht die Bezeichnung

   "mittlere Fluktuation des Windes".

   Es ist die mittlere Veränderung des Windfeldes und nicht die Streuung des Windes um einen Mittelwert.

3.2 Aktueller und geostrophischer Querwind

Die hydrodynamische Grundgleichung ist für den Gebrauch in der Luftfahrt zu folgender Form abgewandelt:

(3.2.1) $$\frac{dp}{dy} = 2 \cdot \omega \cdot \sin\varphi \cdot w_x \cdot \varrho$$

(3.2.2) $$\frac{dp}{dz} = - g \cdot \varrho$$

Es sei:

x	Koordinate senkrecht zum Flugweg
y	Koordinate parallel zum Flugweg
z	Koordinate vertikal
w_x	Windgeschwindigkeitskomponente in X-Achse
dp/dy	horizontaler Gradient des Luftdruckes (p) längs des Flugweges (y)
dp/dz	vertikaler Gradient des Luftdruckes (p)
g	Schwerebeschleunigung der Erde
ϱ	Luftdichte
ω	Winkelgeschwindigkeit der Erde
φ	Geographische Breite

Die Differenz zwischen der absoluten (geometrischen) Flughöhe h_r und der barometrischen Flughöhe h_p (ICAO-Standard-Atmosphäre) wird D-Wert D genannt:

(3.2.3) $$D = h_r - h_p$$

Wird bei konstantem Druck geflogen (h_p = const), ist die Änderung der absoluten Flughöhe zwischen Punkt 1 und Punkt 2 :

(3.2.4) $$\Delta z = D_2 - D_1$$

Aus diesen Gleichungen ist die Geschwindigkeit q des geostrophischen Querwindes zu berechnen. Mit Δy als zurückgelegter Distanz zwischen Punkt 1 und Punkt 2 folgt:

(3.2.5) $$q = \frac{\Delta z}{\Delta y} \cdot \frac{g}{2 \cdot \omega \cdot \sin\varphi}$$

$\Delta z / \Delta y$ ist die Neigung der Isobaren-Fläche;

q ist die Komponente des geostrophischen Windes längs der X-Achse (senkrecht zum Flugzeugkurs) und wird als "fluggeostrophischer Querwind" bezeichnet zur Unterscheidung vom "geostrophischen Wind", der aus dem Isohypsen-Abstand in einer Wetterkarte des 200-mb- oder 300-mb-Niveaus errechnet wird.

Die geometrische Höhe des Flugzeuges wird durch die Anzeige eines Funkechos, welches vom Boden zum Flugzeug zurückkehrt, bestimmt. Über dem Meere entspricht diese Echolotung der wahren Flughöhe. Dieser auch "Radarhöhe" genannte Parameter wurde nicht registriert und mußte deshalb den Navigationslogs (Flug-Protokollen) entnommen werden. Wir können mit den Eintragungen Δz und somit die seitliche flug-geostrophische Abdrift berechnen und sie mit der seitlichen Abdrift aus der Doppler-Radar-Anlage des Flugzeuges vergleichen. Die registrierten Kurven sind mit Planimetern über jene Zeitintervalle integriert worden, für die Radar- und Druckhöhen im Flug-Protokoll vorliegen. Aus den erhaltenen Mittelwerten folgt ein mittlerer Wind für das jeweilige Beobachtungsintervall.

Bei allen Vergleichen zwischen flug-geostrophischen und tatsächlichen Querwinden ist es ein Nachteil, daß die in Kursrichtung liegende Komponente des geostrophischen Windes nicht erfaßt werden kann. Leider ergänzen Messungen anderer im zivilen Luftverkehr eingesetzter Flugzeuge nicht die fehlende Komponente des geostrophischen Windes, weil sich die Flugzeuge auf etwa parallelen Kursen bewegen.

Bild 1 (Seite 49) zeigt die Korrelation zwischen der tatsächlichen Querwind-Komponente (integriert) und der flug-geostrophischen Querwind-Komponente, die aus der barometrischen und der geometrischen Flughöhe berechnet ist. Der Korrelationsfaktor liegt mit 0,73 weit über dem Zufallshöchstwert (0,211). Die gute Verknüpfung der Wertepaare auch an der Grenze zwischen Troposphäre und Stratosphäre läßt eine Auswertung in bezug auf bestimmte Wetterlagen erfolgversprechend erscheinen.

Der Schwerpunkt der in Bild 1 dargestellten Punktwolke liegt im negativen Bereich. Die mittlere Differenz zweier aufeinander folgender D-Wert-Messungen ist also negativ; d.h. es fand im Mittel eine Versetzung des Flugzeuges nach rechts (Steuerbord) statt, weil nach Gleichung (3.2.4) $\Delta z < 0$ wird und nach (3.2.5) auch $q < 0$. Die Atlantik-Überflüge erfolgen auf dem kürzesten Flugweg. Durch diese flugbetriebliche Maßnahme und die zufällige

Zusammenstellung der Flüge sind innerhalb des Kollektivs die Flugzeuge im Mittel mit dem herrschenden Wind zum Kurs nach "rechts" versetzt worden. Der flug-geostrophische Querwind war im Mittel mehr als 4 m sec^{-1} größer als der integrierte aktuelle Wind senkrecht zum Kurs über Grund. Durch die überwiegend zyklonale Krümmung der Isohypsen ist der aktuelle Querwind gegenüber dem Druckgradienten längs des Flugweges im Mittel zu niedrig.

Das Bild 2 (Seite 49) zeigt die Darstellung der Querwind-Komponenten und der Lufttemperatur auf einem Flug der Deutschen Lufthansa mit Registrierungen am 6.4.1962 in 214 mb. Die Tropopause wurde auf ungefähr 35° W durchflogen. Aus diesem Beispiel ist zu erkennen, daß besonders wertvolle Hinweise über die Zirkulation im Tropopausen - Niveau durch eine Registrierung der absoluten Flughöhe und der Druckhöhe des Flugzeuges gewonnen werden können. Einzelheiten zu diesem Flug sind aus Abschnitt 7.1 zu ersehen.

Die gute Korrelation zwischen den Querwind-Komponenten des integrierten Flugzeug-Windes und dem aus Radar- und Druckhöhe berechneten flug-geostrophischen Wind zeigt, daß beide Verfahren meteorologisch wertvolle Informationen liefern. Allerdings erlaubt die Güte der Einzelmessung nur eine statistische Auswertung der flug-geostrophischen Querwind-Komponenten. Ein Vergleich der flug-geostrophischen Querwind-Komponenten mit "spot-winds" (tatsächlicher Wind für einen Ort, im Flugzeug über Doppler-Radar gemessen, nicht integriert) führt auf eine breite Streuung der Wertepaare, weil die Fluktuationen des Windes nicht ausgeglichen sind. Es ist zweckmäßig, wie in dieser Arbeit geschehen, die zur Windberechnung notwendigen Parameter über jene Zeitintervalle zu integrieren, für die zu Beginn und Ende die Druckhöhe und geometrische Flughöhe abgelesen wurde. Diese Zeitintervalle müssen bei den vorhandenen Meßfehlern länger als 10 Zeitminuten sein (Flugstrecke etwa 150 km). Der mittlere Fehler der Einzelmessung beträgt in diesem Fall 10 m sec^{-1}, bei einer Flugstrecke von 300 km nur 5 m sec^{-1}. Die Fehler sind kleiner, sofern keine barometrische Höhenänderung vorgenommen wird (keine Änderung des "Flugniveaus").

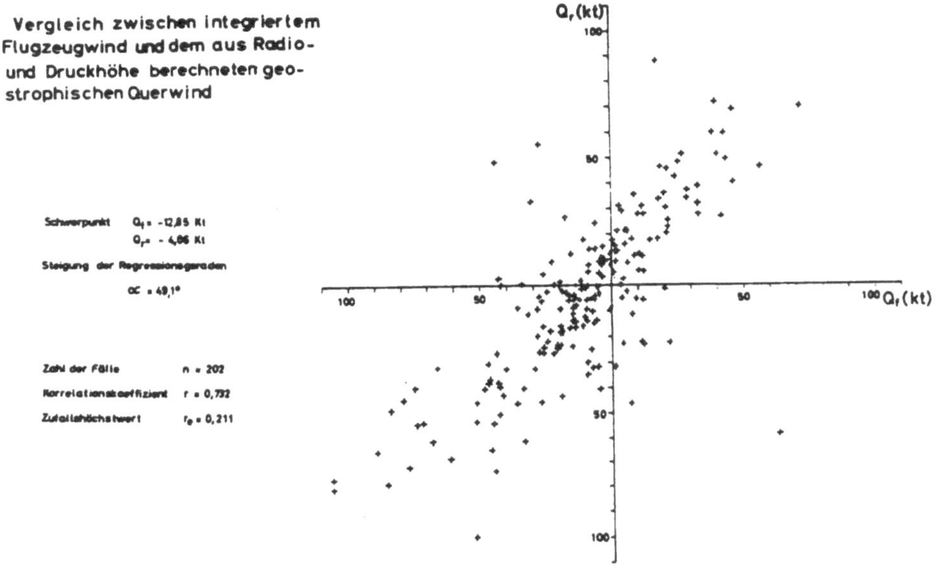

Bild 1 : Korrelation geostrophischer und aktueller Querwind

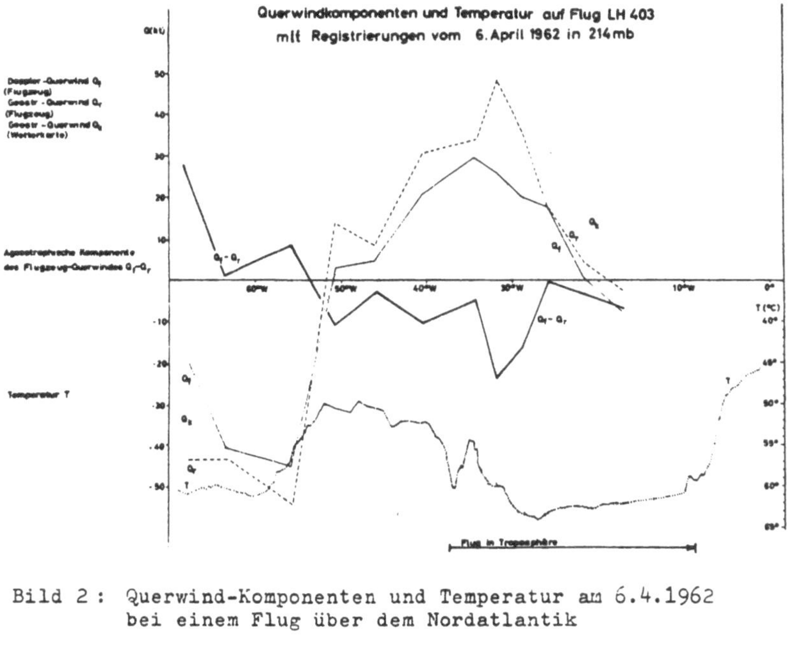

Bild 2 : Querwind-Komponenten und Temperatur am 6.4.1962 bei einem Flug über dem Nordatlantik

3.3 Das mittlere Windfeld und seine Mesostruktur

Die Winddaten, welche mit dem Düsenverkehrsflugzeug D-ABOD der Deutschen Lufthansa vom Typ Boeing 707-430 in den Monaten März 1962 bis Mai 1962 sowie März 1963 bis Oktober 1963 registriert wurden, verteilen sich über dem nördlichen Atlantik zwischen dem 60. und 50. Breitengrad gleichmäßig. Über beiden Kontinenten häufen sich die Daten längs der Flugstraßen, sowie über See längs der nordamerikanischen Küste zwischen Neufundland und New York. Aus beiden Registrierperioden stehen 8968 Einzelwerte (für 1962: 2618 Werte und für 1963: 6350 Werte) zur Verfügung, die auf 143 Flügen erflogen wurden. Dieses anscheinend umfangreiche Material reicht nicht aus, um statistisch einwandfreie Interpretationen vorzunehmen. Es ergeben sich in Teiluntersuchungen Unterschiede zwischen den beiden Jahren, die nur meteorologisch bedingt sein können. Deshalb ist es nicht angebracht, komplizierte, allgemein gültige Aussagen zu versuchen, sie einzuschränken und statistisch zu beweisen. Andere Jahre werden Widersprüche liefern. Es ist besser, überschaubare Prozesse mit Hilfe von Streuungsanalysen und Korrelationsrechnungen nachzuweisen; anderenfalls besteht Gefahr, sich in mathematischen Formalismus zu verlieren und sich von der Realität zu entfernen.

Zur Untersuchung der Mesostruktur der Atmosphäre bieten sich für ein Koordinatensystem drei Orientierungen an:

1. Orientierung nach geographisch Nord und Ost:
2. Orientierung nach Flugzeug, Achsen nach Rumpf- und Tragfläche ausgerichtet;
3. Orientierung nach Luftströmung, Achsen tangential bzw. normal zur Luftströmung ausgerichtet.

Im System 2 können am ehesten Fehlerquellen des Meßsystems nachgewiesen werden (siehe Abschnitt 9). Die Systeme 1 und 2 dienen zur Erfassung meteorologischer Vorgänge.

Die Meßwerte des Flugzeuges wurden für die vorliegende Arbeit für einen zeitlichen Abstand der Meßpunkte von sechs Minuten Flugzeit ermittelt; das bedeutet eine horizontale Distanz von etwa 90 km Luftweg. In beiden Registrierperioden 1962 und 1963 wurde im Mittel fast reiner Westwind erflogen, wie folgende Übersicht zeigt:

Jahr	Komponenten		Mittlere Streuung
1962	Süd: 0,8 m sec^{-1}	West: 14,5 m sec^{-1}	23,0 m sec^{-1}
1963	0,1	15,2	22,3 m sec^{-1}

Der mittlere Fehler für die Einzelmessung des Windes beträgt weniger als 2,9 m sec^{-1}, wie im Abschnitt 9 gezeigt wird.

Bildet man die Differenzen jeweils zweier im Intervall von sechs Minuten Flugzeit benachbarter erflogener Winde, lautet die mittlere Streuung* dieser Differenzvektoren 5,8 m sec^{-1} (für 1963: 6,4 m sec^{-1}). Für die Route Europa - Amerika ist sie mit 5,3 m sec^{-1} (1963: 6,2 m sec^{-1}) gegenüber 6,2 m sec^{-1} (1963: 6,7 m sec^{-1}) für die entgegengesetzte Flugrichtung etwas geringer. Für diese Werte wird im folgenden Text statt des Ausdruckes "mittlere Streuung" der Ausdruck "mittlere Fluktuation" des Windes verwandt, weil es sich nicht um Abweichungen der Einzelwerte vom Mittelwert sondern um örtliche und zeitliche Veränderung des Windes handelt.

Die Flüge auf der Route nach Nordamerika meiden starken Westwind als Gegenwind, weichen also dem Strahlstrom aus; während sie ihn in umgekehrter Richtung meist nutzen können. Es ist zu vermuten, daß der Strahlstrom den Unterschied zwischen den Streuungswerten verursacht. Bei Wahl des geographisch bezogenen Koordinatensystems 1 bleibt der Einfluß der Makrostruktur erhalten; während in dem zur mittleren Luftströmung bezogenen System 3 mesostrukturelle Eigenschaften stärker hervortreten und makrostrukturelle rechnerisch unterdrückt werden.

Rechnet man in einem strömungsbezogenen Koordinatensystem nach Muster 3 , reduziert sich die mittlere Streuung des Differenzvektors nicht, weil der Betrag des Vektors gleich ist, aber die Streuung seiner horizontalen Komponenten läßt Schlüsse auf isotrope Eigenschaften der Turbulenz zu. Zur Errechnung der Streuungswerte im Koordinatensystem 3 wird der Nachfolger eines Windwertes auf diesen projiziert und die Differenz der Komponenten in Richtung des herrschenden Windes und senkrecht dazu gebildet.

Senkrecht zur herrschenden Windrichtung ist die mittlere Fluktuation des Windes etwas geringer derjenigen parallel zur Strömung, nämlich für

1962: Strömungsparallel 4,2 m sec^{-1} Strömungsnormal 3,9 m sec^{-1}
1963: 4,7 m sec^{-1} 4,4 m sec^{-1}

Zwischen den Komponenten der Differenzvektoren besteht tangential bzw. normal zur Strömung nur geringe Korrelation (im Zufälligkeitsbereich).

Diese Ergebnisse stehen in gewissem Widerspruch zu Angaben von S.K.Kao und H.D.Woods (29), die eine Auswahl von Flügen aus dem "Project Jet Stream" der U.S.A. betrachten. In ihrer Arbeit zeigt die "variance" der horizontalen Geschwindigkeiten zwischen strömungsparalleler und -transversaler Richtung merkliche Unterschiede; und zwar ist im Falle stromparalleler Flüge die "variance" der transversalen Geschwindigkeitskomponenten doppelt so groß wie die der longitudinalen Geschwindigkeitskomponenten.

Die Flüge des "Project Jet Stream" wurden nach der Richtung des Strahlstromes ausgelegt. Die höhere Streuung in transversaler Richtung zur mittleren Strömung wird vermutlich durch den seitlichen Anschnitt des Strahlstrom-Kerns und der dabei durchflogenen hohen Windgradienten vorgetäuscht. Kurse auf einer Stromlinie, Isohypse oder gar Isotache können nicht genau eingehalten werden. Andererseits ist der Strahlstrom keinesfalls eine laminare Strömung mit konstanter Richtung und Geschwindigkeit. Mäander und Wirbel im mesostrukturellen Bereich vermögen eine linsen- oder fingerförmige Struktur der Starkwindzonen vorzutäuschen und erzeugen hohe Streuungswerte des Windfeldes. Die Linienflüge der Luftverkehrsgesellschaften längs des Weges minimaler Flugzeit können derartige Strukturen erfassen, weil sie nicht zur speziellen Erkundung der meteorologischen Vorgänge im Strahlstrom erfolgten. Diese Flüge können nicht unterteilt werden in solche, die zum Kern des Strahlstromes senkrecht, und solche, die zum Kern des Strahlstromes parallel verlaufen. Die vektoriellen Winddifferenzen (gebildet nach System 3 zwischen zwei im Abstand von sechs Minuten Flugzeit benachbarten Winden) werden deshalb in vier Gruppen sortiert. Die Gruppen unterscheiden, ob zum Kurs des **Flugzeuges** Gegenwind, Rückenwind, Seitenwind von rechts oder Seitenwind von links herrschte, als die vektorielle Differenz erflogen wurde. Die Ergebnisse dieser Rechnung stellt Tabelle 1 zusammen (Seite 54).

Das Material wurde den Glättungsformeln (Linke (36))

$$\bar{w}_i = \frac{1}{4} (w_{i-1} + 2 w_i + w_{i+1}) \quad ; \tag{3.3.1}$$

$$\bar{w}_i = \frac{1}{25} (5 w_i + 4 (w_{i-1} + w_{i+1}) + 3 (w_{i-2} + w_{i+2}) + 2 (w_{i-3} + w_{i+3}) + (w_{i-4} + w_{i+4})); \tag{3.3.2}$$

$$\bar{w}_i = \frac{1}{125} (25 w_i + 24 (w_{i-1} + w_{i+1}) + 21 (w_{i-2} + w_{i+2}) + 7 (w_{i-3} + w_{i+3}) + 3 (w_{i-4} + w_{i+4}) - 2 (w_{i-6} + w_{i+6}) - 3 (w_{i-7} + w_{i+7})); \tag{3.3.3}$$

sowie der Mittelung

$$\bar{w}_i = \frac{1}{2n+1} \sum_{i=-n}^{i=+n} w_i \qquad \text{für } n = 1, 3, 5, 7, 9 \tag{3.3.4}$$

unterworfen.

Wird der Differenzvektor zwischen einem beobachteten Wind und einem nach Gleichungen (3.3.1) - (3.3.3) geglätteten Strömungsvektor in Koordinaten normal und tangential zur mittleren Windströmung gebildet, ergeben sich mittlere Streuungen nach Tabelle 1 . Die Spalte 1 beschreibt das Kollektiv, die Spalte 2 gibt an , ob der Wert die strömungstangentiale (t) oder die strömungsnormale (n) Komponente charakterisiert. In Spalte 3 (für Jahr 1962) und Spalte 4 (für Jahr 1963) sind die mittleren Fluktuationen des Windes, errechnet aus vektoriellen Winddifferenzen bei einem zeitlichen Abstand von 6 Minuten, eingetragen. In den folgenden Spalten der Tabelle sind Ergebnisse für mittlere Streuungen der vektoriellen Differenzen aufgezeigt, die sich aus dem Vergleich der einzelnen Windwerte mit Windmitteln über Distanzen von 12, 48 und 84 Minuten bei Benutzung der niedergeschriebenen Glättungsformeln ergeben.

Tabelle 1: Mittlere Fluktuation und mittlere Streuung des Windes Angaben in $m\,sec^{-1}$

1	2	3	4	5	6	7	8	9	10
Intervall in min		6		12		48		84	
Windmittel aus Stützstellen		1		3		9		15	
Jahr		1962	1963	1962	1963	1962	1963	1962	1963
Gesamtkollektiv	t	4,0	4,4	1,4	1,5	5,5	5,3	5,4	4,8
	n	3,8	4,3	1,0	1,1	2,2	2,5	2,3	2,6
Flug gegen Strom	t	3,6	3,9	1,3	1,4	5,4	5,1	5,4	4,4
	n	3,3	3,8	1,0	1,1	1,8	2,2	1,9	2,3
Flug mit Strom	t	4,3	4,2	1,5	1,5	5,1	5,1	4,8	5,0
	n	4,2	4,4	1,1	1,2	2,4	2,5	2,5	2,6
Wind von rechts	t	3,8	4,6	1,4	1,6	4,4	5,2	4,1	5,9
	n	3,8	4,7	1,0	1,4	2,2	2,8	2,2	2,9
Wind von links	t	4,2	4,9	1,3	1,5	6,8	5,8	6,6	5,0
	n	3,8	4,2	1,2	1,2	2,6	2,5	2,6	2,6
Flug parallel zum Strom	t	4,0	4,0	1,4	1,4	5,2	5,1	5,1	4,7
	n	3,8	4,2	1,0	1,1	2,1	2,4	2,2	2,5
Flug senkrecht zum Strom	t	4,1	4,8	1,3	1,6	6,1	5,5	5,9	5,0
	n	3,8	4,4	1,1	1,3	2,4	2,7	2,5	2,8

t = Komponente tangential zur mittleren Luftströmung
n = Komponente normal zur mittleren Luftströmung
min = Zeitminuten

Glättungen von Kurven sollten grundsätzlich nur dann vorgenommen werden, wenn bei Vermehrung der Beobachtungszahl die Änderung der Ordinaten der Punkte einer Kurve stetig erfolgt (siehe Baur in Linke (36)). Die in dem groben Gitternetz errechneten Ergebnisse lassen zwar große Zweifel aufkommen, ob diese Annahme für Untersuchungen der Mesostruktur der Atmosphäre richtig ist. Für Vergleiche der erflogenen Winde, deren horizontaler Abstand ungefähr 90 km Luftweg beträgt, mit Wetterkarten, deren Analyse nach den Messungen des synoptisch-aerologischen Beobachtungsnetzes mit einer durchschnittlichen Maschenweite von 300 km (über dem nördlichen Atlantik mehr als 1000 km) erfolgt, ist eine Glättung empfehlenswert.

Die Werte der mittleren Streuung für die vektoriellen Differenzen zwischen erflogenen Winddaten und den nach der Gleichung (3.3.1) geglätteten Mitteln entsprechen ungefähr den Ergebnissen, die S.K.Kao und H.D.Woods (29) durch Auswertung von Flügen des "Project Jet Stream" erhielten. Allerdings ergaben sich im vorliegenden Material, wie aus Tabelle 1 zu erkennen, keine wesentlichen mesostrukturellen Unterschiede zwischen den Streuungen des Wind-Differenzvektors in longitudinaler beziehungsweise transversaler Richtung des Fluges zur Strömung. Die Mesostruktur der Atmosphäre ist im allgemeinen isotrop. Die Ergebnisse von Kao und Woods sind hiernach nicht für die Atmosphäre zu verallgemeinern.

Sehr deutlich tritt das Verhältnis 1 : 2, wie es durch Kao beschrieben wird, zwischen den mittleren Streuungen der normalen und tangentialen Strömungskomponente auf, wenn die Mittelung der Werte über eine Flugzeit von mehr als 30 Minuten (etwa 400 km) erfolgt. Dieses Verhältnis wird durch Eigenschaften der Makrostruktur auf dem Flugweg (Bündelung und Konzentration des Starkwindes in schmale aber langgestreckte Gebiete) erzeugt.

Die nach der Glättungsgleichung (3.3.1) gewonnenen Werte (Tabelle 1, Spalten 5 und 6, Zeitintervall 12 Minuten) zeigen die geringsten Streuungen der Differenzvektoren zwischen aktuellem und mittlerem Wind, weil die Mittelwerte dem lokalen Wind weitgehend rechnerisch angepaßt sind. Die Mittelung über große Distanzen erzeugt einen Mittelwert, der vom Einzelwert stark abweicht.

Die hohen Werte der mittleren Fluktuation des Windes bedeuten eine starke Veränderlichkeit des Windes im Meso-Bereich zwischen 100 und 200 km, wobei die Tendenzen des Feldes über diese Distanz vielfach erhalten bleiben. Feinuntersuchungen des vorliegenden Registriermaterials lassen erkennen, daß der Mesostruktur kleinräumige (mikrostrukturelle) Fluktuationen überlagert sind, die in der Stärke unter 1 m sec^{-1} liegen.

Die Ausdehnung mittelräumiger Strömungsgebilde scheint nach unseren Ergebnissen über 200 km, aber unter der Auflösungsgrenze einer Höhenwetterkarte zu liegen. Es ist vorerst nicht zu klären, ob es sich dabei um Wirbel, Mäander der Starkwindzone, fingerförmige Strombilder oder linsenartige Gebiete mit abweichender Geschwindigkeit handelt. Die Wirbel- oder Strömungssysteme brauchen nicht notwendig in einer Isobarenfläche abgeschlossen zu sein. Sie können die Fläche anschneiden oder sich der herrschenden großräumigen Strömung überlagern. Die etwas höheren Werte mittlerer Streuung und Fluktuation in der zur Strömung tangentialen Komponente deuten unter Berücksichtigung der Messungen von Reiter (51) auf linsenförmige Wirbel im mesostrukturellen Bereich des Windfeldes, wobei die Längsachse in Richtung der großräumigen Strömung liegt.

Nach Tabelle 1 sind große Windunterschiede auf kurze Entfernung im Bereich der oberen Troposphäre und unteren Stratosphäre nach statistischen Gesetzen kaum zu erwarten. Für die Flugsicherheit ist aber nicht ein statistisches Ergebnis sondern der Einzelfall entscheidend. Deshalb werden in Abschnitt 5 extreme Windunterschiede behandelt.

Der erflogene Windvektor ist im Mittel niedriger als der geostrophische Windvektor (aus der Wetterkarte) oder der Gradientwindvektor (aus der Wetterkarte). Im allgemeinen wird ein scharfer Luftdruckgradient zwischen zwei Radiosonden - Stationen beim Analysieren der Wetterkarte ausgeglichen und nicht auf ein schmales Band konzentriert. Falls tatsächlich ein schmales Band hoher Windgeschwindigkeit existiert und dort ein starker, daneben ein schwacher Luftdruckgradient herrscht, wird in den meisten Fällen der erflogene Wind niedriger als der geostrophische sein, welcher der Wetterkarte entnommen wird. Nur in den wenigen Fällen, in denen das Flugzeug im schmalen Starkwindband fliegt, wird die Geschwindigkeit des erflogenen Windes höher als die des geostrophischen Windes oder des Gradientwindes (nach der Wetterkarte) sein.

Auch für das 500-mb-Niveau konnten eigene Untersuchungen (13) zeigen, daß der erflogene Wind im Mittel niedriger als der geostrophische Wind aus der Wetterkarte ist. Die hohen Windgeschwindigkeiten konzentrieren sich dort ebenfalls auf schmale Bänder. Untersuchungen der Deutschen Lufthansa (76) bestätigen diese Ergebnisse, denn die an Hand der Wetterkarten geplanten Flugzeiten für Flüge von Amerika nach Europa waren überwiegend kürzer als die tatsächlich erflogenen.

3.4 Die Mesostruktur des Windfeldes in Abhängigkeit von der Windrichtung

Wir beschränken die Ausführungen auf den mesostrukturellen Bereich. Die Meßergebnisse werden im Bereich unter 10 km Ausdehnung (Mikrostruktur) sehr deutlich von den Flugeigenschaften des Meßträgers beeinflußt. Statistische Untersuchungen zur Makrostruktur schließen sich von selbst aus, weil das vorliegende Material hierfür nicht ausreicht.

Die vektoriellen Differenzen zweier benachbarter Windwerte des vorliegenden Sechs - Minuten - Gitternetzes aus den Flugzeugregistrierungen sind nach der sechsunddreißigteiligen Windrose sortiert. Für jedes der 36 Kollektive sind Rechnungen in geographisch- und strömungsbezogenen Koordinatensystemen durchgeführt worden. Die Ergebnisse über das strömungsbezogene System werden in Bild 3 dargestellt.

In Bild 3 trägt die Abszisse die Windrichtung und die Ordinate die Werte der mittleren Streuung (bzw. mittlere Fluktuation) in $m\ sec^{-1}$. Die oberen Kurvenzüge verbinden die Werte der mittleren Streuung für die Komponenten (in tangentialer und normaler Richtung zur Strömung) der vektoriellen Differenzen zwischen erflogenen und geostrophischen Winden aus einer im Deutschen Wetterdienst gebräuchlichen Arbeitswetterkarte. Die unteren Kurven verknüpfen die Werte der mittleren Fluktuation für die Komponenten (in tangentialer und normaler Richtung zur herrschenden Strömung) der vektoriellen Differenzen jeweils benachbarter Windwerte.

Das Bild 3 zeigt für die östlichen Windrichtungen einen unruhigeren Kurvenverlauf als für westliche Richtungen. Die mittleren Fluktuationen schwanken um $4\ m\ sec^{-1}$. Im südöstlichen und südwestlichen Bereich liegen die Werte für die mittlere Fluktuation der strömungstangentialen Komponente um mehr als $0{,}5\ m\ sec^{-1}$ über denjenigen der strömungsnor-

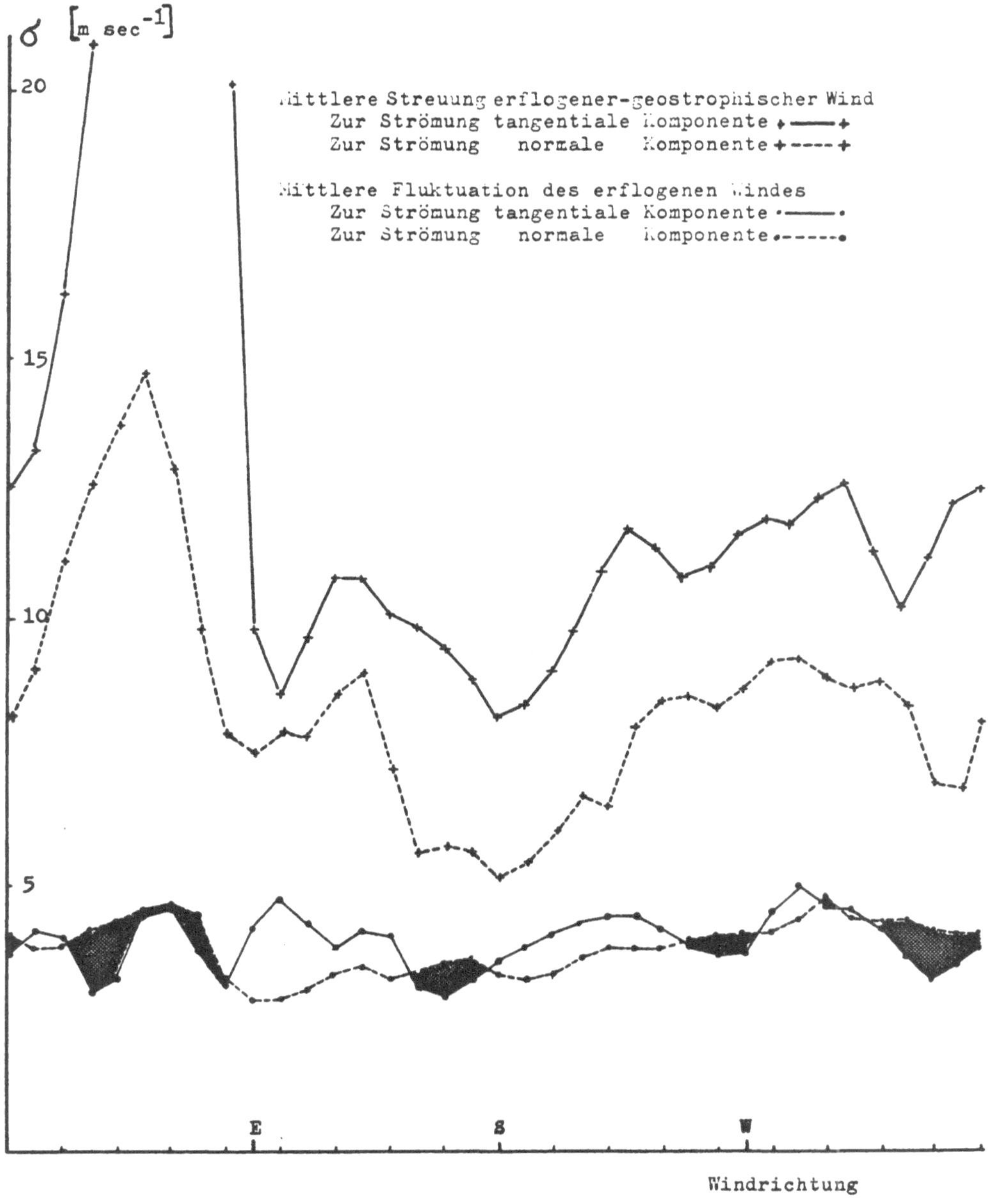

Bild 3: Mittlere Fluktuation der Windes und mittlere Streuung des Differenzvektors erflogener-geostrophischer Wind in Abhängigkeit von der mittleren Windrichtung

malen Komponente. Für die Kollektive mit südlichen, vor allem nordöstlichen bis nordwestlichen Winden liegen die Werte der mittleren Fluktuation für die Komponente normal zur herrschenden Strömung etwas höher (Raster in Bild 3). Die Luftteilchen scheinen bei westlichen Winden ellipsenförmige Störungsbahnen zu durchlaufen. Wirbelartige "Störungen", die einer großräumigen Strömung überlagert sind, können dieses Ergebnis erzeugen. (Die westlichen Winde sind im Mittel höher als östliche oder nördliche). Die erflogenen östlichen Winde stammen meist aus hochreichenden Tiefdruckgebieten.

Die Befunde bestätigen die synoptische Erfahrung, daß östliche Winde in ihrer Richtung weniger beständig als westliche sind und sich schwer über große Strecken durchsetzen. Das darf aber nicht dazu verleiten, die Existenz einer östlichen Höhenströmung auf dem nördlichen Atlantik über mehrere Tage zu leugnen. Die hohen Abweichungen zwischen geostrophischen und erflogenen Winden sowie die dadurch bedingten hohen Werte der mittleren Streuung in den Kollektiven des nordöstlichen Sektors sind durch Analysenfehler der Wetterkarte bedingt, die über mehrere Tage hinweg nicht bemerkt wurden. In jenen Tagen bildete sich eine östliche Luftströmung über dem nordöstlichen Atlantik mit Geschwindigkeiten über 30 m sec^{-1} so ungünstig aus, daß sie mit dem bestehenden Radiosonden-Netz nicht erfaßt wurde.

Die Abweichungen des erflogenen Windes vom geostrophischen Wind werden in folgendem System erläutert:

1. Der ageostrophische Windvektor ist dem geostrophischen Wind zu addieren, um den erflogenen Wind zu erhalten.

2. Das strömungsbezogene Koordinatensystem ist für diesen Teil der Untersuchungen nach dem geostrophischen Wind ausgerichtet. Seine t-Achse (tangential) ist die Richtung des geostrophischen Windes, n-Achse liegt dazu normal.

3. Die Winddaten sind nach der erflogenen Windrichtung in Kollektive selektiert.

Die mittlere Streuung der strömungsparallelen ageostrophischen Komponente ist wesentlich höher als die dazu normal liegende. Auffallend niedrig ist die mittlere Streuung in den Kollektiven mit südlichen Winden. Diese Werte dürften bei guter Analyse und mehr Meßdaten allgemein zu erreichen sein. Die Grenze einer Verbesserung liegt bei einem Streuungswert über $4\,m\,sec^{-1}$. Der Wert ist durch die mittlere Fluktuation des Windes im mesostrukturellen Bereich vorgegeben.

Die festgestellten Differenzen zwischen geostrophischen und erflogenen Winden setzen sich aus echten ageostrophischen Beiträgen und Wetterkarten-Fehlern zusammen. Deshalb sind die hohen Werte für die mittlere Streuung der vektoriellen Differenzen zwischen erflogenen und geostrophischen Winden in den Kollektiven mit südwest- und westlichen Winden nicht zu verallgemeinern. Sie werden hauptsächlich durch Unterschiede bei hohen Windgeschwindigkeiten im Bereich des Strahlstromes verursacht. Wie in den folgenden Abschnitten 3.5 und 3.6 gezeigt wird, bleibt die genaue Lage der Strahlstromachse unbekannt. In einer Höhenwetterkarte kann deshalb weder die Richtung noch der Abstand der Isohypsen genau gezeichnet werden.

Der Gradientwind bringt gegenüber dem geostrophischen Wind im Vergleich mit erflogenen Winden keine Differenzminderung. Die mittleren Streuungen der vektoriellen Differenzen zwischen erflogenen Winden und Gradientwinden sind für die Kollektive mit südöstlichen und mit west- bis nordwestlichen Winden sogar höher als diejenigen im Vergleich mit geostrophischen Winden. Die ungenaue Zeichnung der Isohypsenkrümmungen, die schlechte Erfassung der Strombahnkrümmungen sowie die Wanderung und Veränderung der Druckgebilde verursachen höhere Abweichungen.

Die Genauigkeit der Wetterkarte wird im Abschnitt 4 unter Berücksichtigung der soeben beschriebenen Merkmale diskutiert.

3.5 Die Mesostruktur des Windfeldes in Abhängigkeit von der Windgeschwindigkeit

Die mittlere Fluktuation des Windes nimmt mit der Windgeschwindigkeit nur wenig zu, und zwar

von 3,5 m sec^{-1} bei einer Windgeschwindigkeit von 5 m sec^{-1}
auf 5,5 m sec^{-1} bei einer Windgeschwindigkeit von 50 m sec^{-1}.

Die Werte der mittleren Fluktuation der strömungsparallelen Komponente liegen bis etwa 30 m sec^{-1} Windgeschwindigkeit wenig über denjenigen der strömungsnormalen Komponente; für höhere Windgeschwindigkeiten übersteigen die Werte der strömungsparallelen Komponente um mehr als 0,5 m sec^{-1} die andere Komponente. Da die Zahl der Messungen mit hohen Windgeschwindigkeiten für statistische Auswertungen meteorologischer Erscheinungen nicht ausreicht, sind die in Bild 4 angegebenen Daten oberhalb von 45 m sec^{-1} nicht als endgültig anzusehen.

Für 1962 liegen die Werte für die mittleren Fluktuationen bei allen Geschwindigkeiten ungefähr 0,5 m sec^{-1} niedriger als 1963. Ihr Verlauf ist gleichartig. Diese Angaben geben einen Hinweis über die Zuverlässigkeit.

Die Zunahme der mittleren Fluktuation des Windes mit der Windgeschwindigkeit ist in der oberen Troposphäre und unteren Stratosphäre gegenüber derjenigen in den bodennahen Schichten nur gering. Dieser Befund überrascht, soweit nur Erfahrungen aus der Technik und der bodennahen Reibungsschicht vorliegen. Die atmosphärische Turbulenz unterscheidet sich von den in der Technik vorkommenden Turbulenzformen grundlegend durch Weiträumigkeit, Rotation der Atmosphäre und wetterhafte Erscheinungen (Raethjen (45 , 47)). Die atmosphärischen Turbulenzen entstehen nicht nur an der Erdoberfläche durch Reibung sondern auch in der freien Atmosphäre durch Austauschvorgänge. Insbesondere greift die " freie Atmosphäre " durch Impulszufuhr von oben in die Vorgänge der bodennahen Reibungsschicht ein (Fall-böen). Damit treten neben die beständig wirkenden Reibungskräfte des Bodens wetterhafte Kräfte aus der freien Atmosphäre.

In der Flughöhe des Düsenluftverkehrs sind für einen Ort die Störimpulse in einer Luftströmung nicht kontinuierlich. Die Störimpulse der freien Atmosphäre sind wetterhaft, konvektiv oder orographisch bedingt. Sie treten deshalb auf der Erde nicht gleichmäßig auf. Orographisch hervorgerufene Störungen sind zwar an den Ort gebunden, ihre Stärke und Wirkung sind aber je nach der Wetterlage (Richtung der Strömung gegen ein Gebirge) verschieden. Konvektiv bedingte Störungen häufen sich je nach Wetterlage in bestimmten Gebieten. Sie wirken allgemein schwächer als orographische Hindernisse in gleicher Höhe, weil ein vom Erdboden konvektiv aufsteigendes Luftpaket von der umgebenden horizontal strömenden Luft beschleunigt wird. Wirksam für einen Störimpuls ist nur der Geschwindigkeitsunterschied zwischen der ungestörten Strömung und dem aufsteigenden Luftpaket. Deshalb ist für die freie Atmosphäre eine geringere Zunahme der mittleren Fluktuation mit der Windgeschwindigkeit festzustellen als in der bodennahen Reibungsschicht.

Zur Beurteilung der Höhe und Zunahme der mittleren Fluktuation des Windes ist auch die "räumliche Größenordnung" zu beachten, in welcher die Winddaten erflogen wurden. Der räumliche Abstand zweier benachbarter Winddaten beträgt ungefähr 100 km. Dadurch werden in einem Meso-Bereich Fluktuationen des Windes erschlossen, die durch Radiosonden nicht mehr erfaßt werden. Die erflogenen Fluktuationen des Windes unterscheiden sich aber von Mikro-Fluktuationen, die an einem Ort gewonnen wurden, weil zwischen zwei erflogenen Wind-Daten neben dem zeitlichen Unterschied auch ein räumlicher Abstand besteht.

Durch den räumlichen Abstand zwischen zwei Windwerten erhöhen sich die Werte der mittleren Fluktuationen des Windes. Die Fälle mit niederen Windgeschwindigkeiten sind (durch die Streckenführung der Flüge) vor allem in Trögen und Allschichten-Zyklonen und weniger in ausgedehnten Höhen-Hochdruckgebieten erflogen worden. Durch die kleine räumliche Ausdehnung der windstillen Gebiete in den Tiefdruckgebilden liegen besonders für niedrige Windgeschwindigkeiten die erflogenen Fluktuationen höher, als mit einem ortsfesten Meßsystem festgestellt würde.

Für alle aus den Jahren 1962 und 1963 vorliegenden Meßserien wurden Autokorrelationsfaktoren berechnet und zwar für die Windwerte und für die vektoriellen Differenzen des Windes zwischen zwei benachbarten Winddaten (horizontaler Abstand 6 Minuten Flugzeit, ungefähr 80 - 100 km). Der Abstand der autokorrelierten Wertepaare variiert zwischen 50 km und 2000 km. Außerdem wurde die Zahl der Wendepunkte für die Windregistrierung jedes Fluges in der Nordkomponente und in der Ostkomponente festgestellt und mit der mittleren Fluktuation des Windes verglichen. Nur in wenigen Fällen (bei Durchflug schmaler Tröge oder Rücken) liegen die Autokorrelationsfaktoren auch für größeren Abstand der Wertepaare außerhalb des Zufallsbereiches. Diese Untersuchungen weisen keine Erhöhung der mittleren Fluktuationen des Windes durch die Verteilung der durchflogenen Hoch- und Tiefdruckgebiete für höhere Windgeschwindigkeiten nach.

Das geostrophische Windfeld zeigt mit zunehmender Windgeschwindigkeit größere Abweichungen vom erflogenen Windfeld. Die vektoriellen Differenzen zwischen geostrophischen Winden aus der Wetterkarte und erflogenen Winden werden in einem nach der geostrophischen Strömung ausgerichteten Koordinatensystem errechnet und nach der Geschwindigkeit des erflogenen Windes geordnet. Die mittlere Streuung der strömungsparallelen Komponente des Differenzvektors zwischen geostrophischen und erflogenen Winden ist für alle Windgeschwindigkeiten um mehr als 1 m sec^{-1} höher als diejenige der strömungsnormalen Komponente. Sie liegt unter 30 m sec^{-1} Windgeschwindigkeit um 10 m sec^{-1} und nimmt für höhere Windgeschwindigkeiten auf Werte um 25 m sec^{-1} zu. Die mittlere Streuung der strömungsparallelen Komponente wächst von 3,3 m sec^{-1} bei 5 m sec^{-1} Windgeschwindigkeit auf 9,0 m sec^{-1} bei 35 m sec^{-1} Windgeschwindigkeit . Für höhere

Windgeschwindigkeiten treten wesentlich größere Abweichungen auf, weil die Lage der Strahlstromachse meist nicht richtig analysiert wird.

Der Gradientwind zeigt gegenüber dem geostrophischen Wind keine Verbesserungen im Vergleich mit dem erflogenen Wind. Grundsätzlich ist der Gradientwind keinesfalls schlechter geeignet als der geostrophische Wind. Lediglich durch den weiten Abstand der Radiosondenstationen über dem Atlantik sind die Isohypsen-Krümmungen falsch gezeichnet, wodurch höhere Abweichungen zu den erflogenen Winden erzielt werden. Die erflogenen Winde charakterisieren mesostrukturelle Eigenschaften der Atmosphäre, während Gradientwinde und geostrophische Winde aus der Makrostruktur des Feldes errechnet werden.

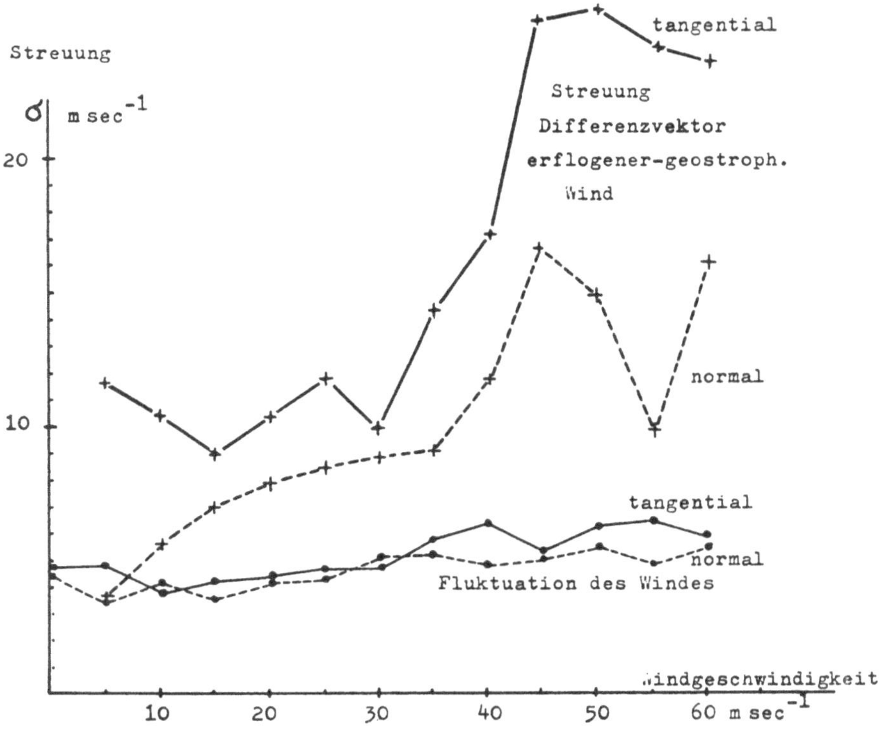

Bild 4: Mittlere Fluktuationen des Windes und mittlere Streuungen des Differenzvektors erflogener - geostrophischer Wind in Abhängigkeit von der Windgeschwindigkeit

3.6 Die Mesostruktur des Windfeldes in Beziehung zur Tropopause und zum jetstream

3.6.1 Beschreibung der Unterteilung

Das vorliegende Material ist in Abhängigkeit zu seiner Lage vom Strahlstrom und von der Tropopause durch ein System von je zehn Ziffern für die Vertikale und die horizontalen Zuordnungen parallel und normal zur Achse des Strahlstromes geordnet worden. Es entstehen dadurch 999 Teilkollektive und ein Gesamtkollektiv, in welchem zur Rechnungskontrolle alle Teilsummen zusammengefaßt sind. Die gebildete dreidimensionale Matrix trägt als Indizes die Zuordnungsziffern 0 bis 9 je nach Lage des Meßpunktes zur herrschenden Strömung und zur Tropopause.

Es wird angenommen, daß die Meßpunkte in einer s-förmig gekrümmten Strömung liegen. Von Norden nach Süden liegen in dieser Strömung der polare und der subtropische Zweig des jetstream. Diese beiden Zweige können in der Mitte des gedachten s-förmigen Bandes vereinigt auftreten (dieser Fall wurde durch Kollektiv 8 berücksichtigt; siehe Erläuterung zu "Quer-Skala", Seite 69). Eine blätterige Struktur der Tropopause wird ausgeschlossen. Eine klare Entscheidung zwischen troposphärischen und stratosphärischen Abschnitten ist für alle Flüge zu treffen.

Sowohl die Windvektoren als auch die vektoriellen Differenzen zweier benachbarter Windwerte (Abstand 6 Minuten Flugzeit) und die vektoriellen Differenzen zwischen erflogenen und geostrophischen Winden wurden in den noch näher zu beschreibenen Systemen geprüft. Zweckmäßigerweise wurden alle drei im Abschnitt 3.3 beschriebenen Koordinatensysteme verwandt:

1. geographisch bezogen, Nord - Ost ;
2. Flugzeug-bezogen, Rumpf - Tragfläche ;
3. strömungsbezogen, strömungsnormal - strömungsparallel.

Die Ergebnisse nach System 1 und 3 werden mitgeteilt.

Die Daten wurden in einer Voruntersuchung nach den jeweils über dem Nordatlantik herrschenden Typen des jetstream folgenderweise unterschieden:

0 kein jetstream zu lokalisieren
1 schwacher jetstream polaren oder subtropischen Typs
2 arktischer jetstream vorherrschend
3 polarer jetstream vorherrschend
4 subtropischer jetstream vorherrschend
5 starker jetstream, polare und subtropische Zweige getrennt
6 starker jetstream, polare und subtropische Zweige vereint
7 polarer jetstream mit östlicher Richtung
8 subtropischer jetstream mit östlicher Richtung

Die Werte, welche sich aus dieser Aufspaltung ergaben, zeigten keine signifikanten Unterschiede. Die Fluktuationen des Windes sind nicht durch den Typ des jetstream, welcher zur Zeit der Messung auf dem Längengrad vorherrschend ist, in ihrer Höhe bestimmt. Die Entscheidung, ob ein Strahlstrom polaren oder subtropischen Ursprungs sei, erfolgte nach der Zirkulation auf der gesamten Nordhemisphäre.

Zur Ordnung der Daten werden drei Skalen eingeführt.

Skala 1 (Vertikal-Skala):

0 keine Zuordnung möglich

Flug in Troposphäre:

1	Flugniveau mehr als 1000 m				unter der Tropopause
2	Flugniveau	1000 m	bis	500 m	unter der Tropopause
3	Flugniveau	500 m	bis	200 m	unter der Tropopause
4	Flugniveau	200 m	bis	0 m	unter der Tropopause

Flug in Stratosphäre:

5	Flugniveau	0 m	bis	200 m	über der Tropopause
6	Flugniveau	200 m	bis	500 m	über der Tropopause
7	Flugniveau	500 m	bis	1000 m	über der Tropopause
8	Flugniveau mehr als 1000 m				über der Tropopause

9 Gesamtkollektiv (Addition aller Werte zur Kontrolle)

Die Höhe der Tropopause wurde den Tropopausenkarten der Flugwetterwarte Frankfurt entnommen, die nach Messungen der Radiosonden analysiert worden sind. Die Entscheidung zwischen den

Kollektiven 4 und 5 fiel für einen Wert nach der im Flugzeug registrierten Lufttemperatur unter Berücksichtigung aller übrigen atmosphärischen Parameter. Die oben angegebenen Maße sind linear-interpoliert der Tropopausenkarte entnommen worden. Sie haben für die Natur nur die eingeschränkte Bedeutung:

Kollektiv 1 sinngemäß "weit in der Troposphäre",
Kollektiv 2 sinngemäß "in der Troposphäre",
Kollektiv 3 sinngemäß "in der Troposphäre unter der Tropopause",
Kollektiv 4 sinngemäß "in der Troposphäre an der Tropopause",

entsprechend die Kollektive 5 bis 8 für die stratosphärischen Werte.

Die horizontale Zuordnung der Meßpunkte in das s-förmig gedachte Stromfeld mit Strahlstrom erfolgte nach 200-mb-Wetterkarten der Flugwetterwarte Frankfurt (Beschreibung Skala 2 und **Skala** 3 auf Seite 69). Die Analysen sind nachträglich durch die Ergebnisse der Registrierungen verbessert worden. Die Zuordnung zum Strahlstrom berücksichtigt nicht die Definition der World Meteorological Organization (WMO), die eine willkürliche untere Grenze der Geschwindigkeit entlang der Achse eines Strahlstromes mit 30 m sec^{-1} festsetzte. Auch die verlangten horizontalen und vertikalen Geschwindigkeitsgradienten blieben unbeachtet, um nicht durch bürokratische Vorschriften physikalische Zusammenhänge zu übersehen. In dieser Arbeit wurde vielmehr von der Vorstellung einer um das polare Tiefdruckgebiet in sich geschlossenen Zirkulation ausgegangen, die einen Frontalstrom mit Maander bildet. Die Linie stärkster Windgeschwindigkeit (nach verbesserter Wetterkarte) zwischen Polarkalotte und Subtropen wird als Achse eines Strahlstromes, bei Auftreten mehrerer Starkwindbänder als Ast eines Strahlstromes (arktisch, polar, subtropisch) definiert. (Es ist dabei gleichgültig, ob die Windgeschwindigkeit über 30 m sec^{-1} liegt). Falls eine derartige Linie nicht zu finden ist, erfolgt Zuordnung in Kollektiv 0. Das Kollektiv 9 dient zur Überprüfung der Rechnung.

Es ist zu beachten, daß Skala 2 (Quer-Skala) eine statistische Unterteilung wiedergibt. Deshalb sind in den Bildern zu Abschnitt 3.6.3 nur die zusammengehörigen Werte aus dem Bereich des polaren Strahlstromes und diejenigen aus dem Bereich des subtropischen Strahlstromes miteinander verbunden worden.

Zur Stromrichtung des jetstream normal liegt

Skala 2 (Quer-Skala):

0 keine Zuordnung möglich

1 links vom
2 im polaren Ast des Strahlstromes
3 rechts vom

4 zwischen polaren und subtropischen Zweigen des Strahlstromes

5 links vom
6 im subtropischen Ast des Strahlstromes
7 rechts vom

8 im Kern eines Strahlstromes mit vereinigten subtropischen und polaren Zweigen

9 Gesamtkollektiv (Addition aller Werte zur Kontrolle)

"rechts" bedeutet: Mit dem Wind im Rücken in Richtung der Strömung vorwärts blickend rechter Hand;
"links" bedeutet: Entsprechend wie oben, aber linker Hand.

Zur Stromrichtung des jetstream parallel liegt

Skala 3 (Parallel-Skala):

0 keine Zuordnung möglich

1 Anfang
2 Mitte zyklonaler Strömungsbogen
3 Ende

4 Wendepunkt zyklonaler - antizyklonaler Strömungsbogen

5 Anfang
6 Mitte antizyklonaler Strömungsbogen
7 Ende

8 Wendepunkt antizyklonaler - zyklonaler Strömungsbogen

9 Gesamtkollektiv (Addition aller Werte zur Kontrolle)

Die Bilder des Abschnitts 3.5 zeigen in der Ordinate jeweils die Kennziffer des zugehörigen Kollektivs (Skala). Die beiden oberen Darstellungen eines Bildes geben in der Abszisse die mittleren Fluktuationen (Streuungen) für die strömungstangentiale und strömungsnormale Komponente der vektoriellen Differenzen, die untere Darstellung gibt dagegen den mittleren Betrag der vektoriellen Differenzen wieder.

3.6.2 Statistische Betrachtung der Fluktuationen und ageostrophischen Komponenten des Windfeldes in vertikaler Richtung (Vertikal-Skala)

Die Ergebnisse der Rechnungen in den verschiedenen Koordinatensystemen zeigen meteorologisch keine Unterschiede. Die Nordkomponente des Windes ist für die Troposphäre größer als für die Stratosphäre. Eine stärkere südliche Komponente des Windes herrscht oberhalb der Tropopause. Die höchsten Werte in der West - Komponente des Windes sind in der Nähe der Tropopause festzustellen, wobei das Maximum der West - Wind - Komponente unmittelbar über der Tropopause liegt. Das Maximum der skalaren Windgeschwindigkeit befindet sich dagegen unter der Tropopause. Die hohen Beträge der Windgeschwindigkeit setzen sich aber im Mittel von der Tropopause weiter in die Stratosphäre als in die Troposphäre hinein fort. Bild 5 zeigt die mittleren Geschwindigkeiten des Windes in Abhängigkeit vom Abstand zur Tropopause. Das Koordinatensystem ist ausnahmsweise nach geographisch Nord und Ost ausgerichtet.

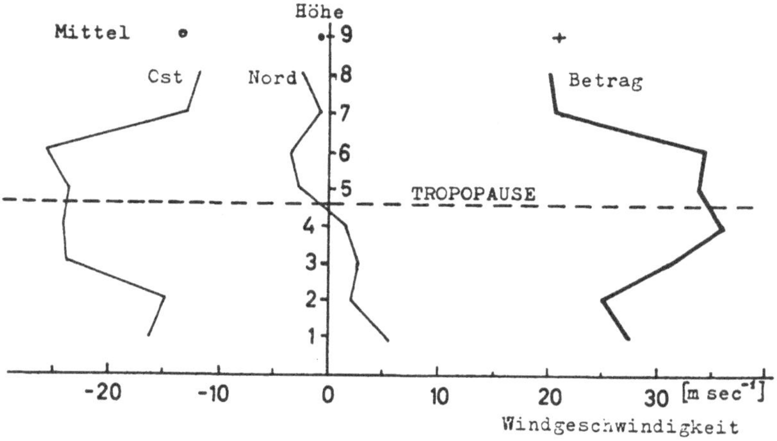

Bild 5 : Mittlere Windgeschwindigkeit in Abhängigkeit zum Abstand von der Tropopause

Die mittleren Windgeschwindigkeiten liegen in der oberen Troposphäre (Kollektive 1 und 2 in der Vertikal-Skala) um etwa 5 m sec^{-1} über denjenigen in der unteren Stratosphäre (Kollektive 7 und 8 in der Vertikal-Skala). Dieser Unterschied könnte durch flugbetriebliche Gründe (Flug auf dem Wege kürzester Flugzeit) verursacht sein. Der Flug Europa - Amerika erfolgt möglichst bei geringen Westwindgeschwindigkeiten und nördlicher als der Flug Amerika - Europa, der südlicher und möglichst mit hohen Westwindgeschwindigkeiten durchgeführt wird. Dadurch wird auf der Strecke nach Amerika eher in der Stratosphäre geflogen als im Gegenflug.

Bild 6 (Seite 72) zeigt die mittleren Fluktuationen des Windes. In der Ordinate bezeichnen die Ziffern Kollektive, welche zur Trennung der Daten gebildet wurden (troposphärische Werte unten, Tropopause liegt zwischen 4 und 5, nach Vertikal-Skala). In der Abszisse sind die mittleren Fluktuationen des Windes aufgetragen. Die mittleren Fluktuationen zeigen die höchsten Werte an der Tropopause. Ihre Abnahme erfolgt nahezu symmetrisch zu dieser Grenzfläche. Die mittlere Fluktuation des Windes ist für die Komponente normal zur herrschenden Strömung an der Tropopause mit 6 m sec^{-1} um 2 m sec^{-1} höher als diejenige in der unteren Stratosphäre und oberen Troposphäre. Dagegen zeigt die Komponente, welche in tangentialer Richtung zur Strömung liegt, lediglich die geringe Zunahme um 0,5 m sec^{-1} in einem Bereich über der Tropopause.

Die im isobaren Flug an der Tropopause auftretenden Windunterschiede scheinen nach dem statistischen Sachverhalt durch Querzirkulationen zur Grundströmung zu entstehen. Die überregionale Strömung bleibt an der Tropopause von den mesostrukturellen Vorgängen anscheinend unbeeinflußt. Das großräumige Druckfeld bestimmt wesentlich die Grundströmung, deren Werte an der Tropopause keine großen Schwankungen (beschrieben durch die strömungstangentiale Komponente der vektoriellen Winddifferenzen) anzeigen. Die Grundströmung kann aber durch die überlagerten Querströmungen (beschrieben durch die strömungsnormale Komponente der vektoriellen Winddifferenzen) großräumig verändert werden, sofern diese über

Skala 1

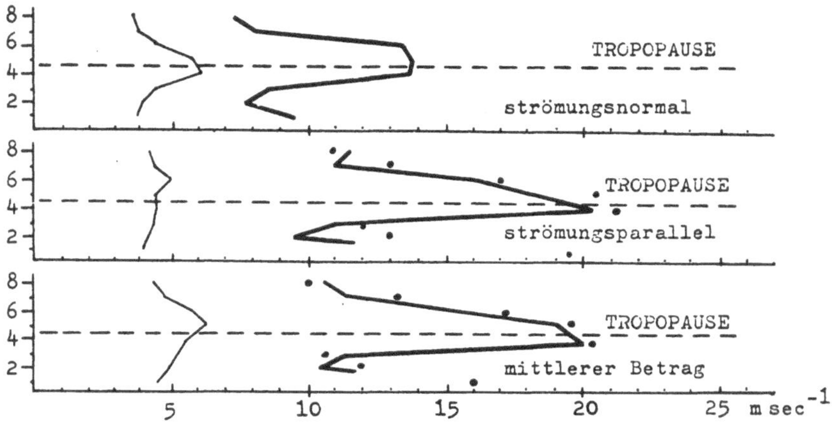

Ordinate: Kennziffer der Kollektive
Abszisse: Mittlere Streuung der Differenzvektoren
geostrophischer-erflogener Wind
Gradientwind - erflogener Wind
Mittlere Fluktuationen des erflogenen Windes

Bild 6: Mittlere Fluktuationen des Windes in Abhängigkeit zum Abstand von der Tropopause (Vertikal-Skala)

längere Zeiten gleichsinnig wirken. Nur in wenigen Fällen überflog das Flugzeug vor dem amerikanischen Kontinent nach wenigen Stunden die gleichen Orte (dann in einem anderen Flugniveau), so daß zeitliche Veränderungen des Windfeldes nicht entdeckt werden konnten.

Der erflogene Wind erscheint als resultierender Vektor verschiedener Felder, nämlich der Grundströmung, hauptsächlich bestimmt durch das großräumige Druckfeld, und einer lokalen, stark veränderlichen Querströmung. Falls die vektoriellen Differenzen zwischen erflogenen und geostrophischen Winden (der Wetterkarte entnommen) nicht gegen den erflogenen Windvektor, sondern gegen den geostrophischen Windvektor projiziert werden, zeigt die strömungsparallele Komponente des Differenzvektors an der Tropopause ebenfalls höhere Streuungen. Bild 6 gibt die mittleren Streuungen der vektoriellen Differenzen zwischen erflogenen und geostrophischen Winden mit Projektion auf die geostrophischen Winde wieder.

Man ist geneigt, die Abweichungen von der Grundströmung als Störungen abzutun oder einfach als Turbulenz zu deuten. Diese "Störungen" werden, ähnlich wie bei der "Allgemeinen Zirkulation", notwendig zur Erhaltung oder Veränderung der Grundströmung beitragen. Jedes Luftmassenelement, dessen Gleichgewicht im geostrophischen Feld gestört ist, vollführt eine Störbewegung auf antizyklonaler Bahn. Da die Windunterschiede an der Tropopause größer als in der übrigen freien Atmosphäre zwischen der 200-mb- und 300-mb-Fläche sind, könnten auch durch die Tropopause Störbewegungen ausgelöst werden, die zu einer Umstellung der Höhenströmung führen.

Die mittlere Streuung für die vektoriellen Differenzen zwischen erflogenen und geostrophischen Winden (Bild 6 fett gezogene Linien) beziehungsweise Gradientwinden (Bild 6 durch Punkte markiert) ist an der Tropopause am größten. Die Werte des Gradientwindes liegen etwas ungünstiger als diejenigen des geostrophischen Windes. Da der erflogene Wind auf den geostrophischen Windvektor projiziert wurde, sind natürlich die Streuungswerte in der Strömungsnormalen für die Differenzvektoren erflogener - geostrophischer Wind beziehungsweise Gradientwind gleich.

Die Werte der mittleren Streuungen für die Differenzvektoren zwischen erflogenen und geostrophischen Winden liegen in der strömungsparallelen Komponente höher als in der strömungsnormalen Komponente. Die Streuungen in der strömungsparallelen Komponente schwanken auch stärker zwischen den einzelnen Unterkollektiven (0 bis 8) als diejenigen in der strömungsnormalen Komponente. Die ungenaue Kenntnis der Lage der Strahlstrom-Achse und die Ungewißheit über die straffe Bündelung des Starkwindfeldes führen zu größeren Abweichungen in der strömungstangentialen Komponente der vektoriellen Differenzen zwischen erflogenen und geostrophischen Winden.

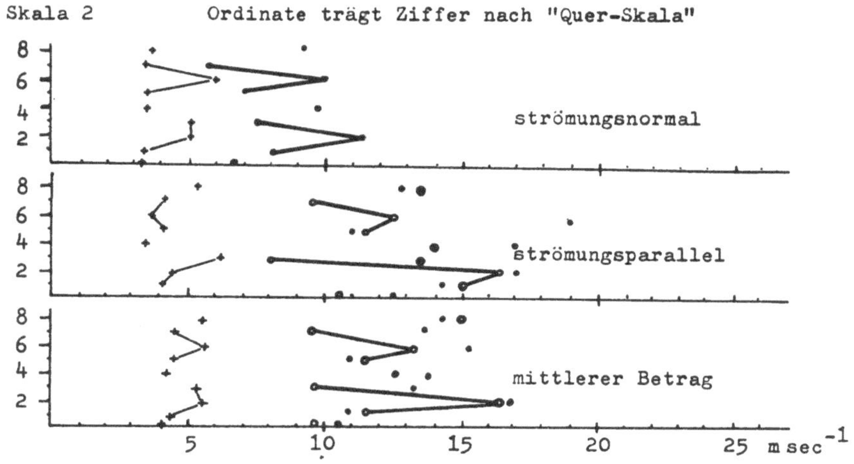

Bild 7: Mittlere Fluktuationen des Windes und mittlere Streuungen der vektoriellen Differenzen zwischen geostrophischen und erflogenen Winden in "Quer-Skala"

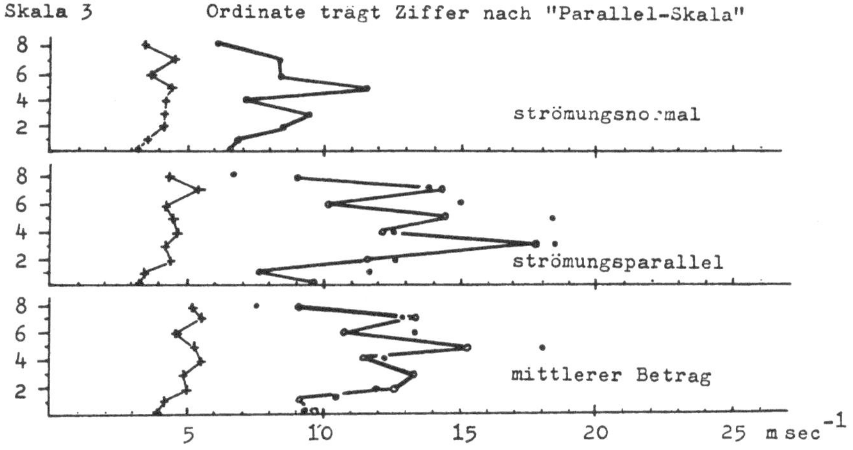

Bild 8: Mittlere Fluktuationen des Windes und mittlere Streuungen der vektoriellen Differenzen zwischen geostrophischen und erflogenen Winden in "Parallel-Skala"

Erklärung: Ordinate Kennziffer der Kollektive nach Skalen 2 - 3
Abszisse mittlere Streuung der Differenzvektoren
━━ o o geostrophischer - erflogener Wind
● ● Gradientwind - erflogener Wind
—— + Mittlere Fluktuationen des erflogenen Windes

3.6.3 Statistische Betrachtung der Fluktuationen und ageostrophischen Komponenten des Windfeldes in der "Quer-Skala"

Die Quer-Skala liegt normal zur Achse des jetstream (Erklärung auf Seite 69). Die höchsten mittleren Beträge der vektoriellen Differenzen zweier benachbarter Windwerte (Abstand 6 Minuten Flugzeit) sind in den Strahlströmen zu finden (Seite 74, Bild 7 unterer Teil). Die mittleren Fluktuationen des Windes zeigen für die Bereiche der Strahlströme ebenfalls höhere Werte, jedoch verhalten sich die Komponenten tangential und normal zur herrschenden Strömung nicht gleichartig. Die mittleren Fluktuationen des Windes sind im Bereich eines jetstream nicht höher in der strömungstangentialen Komponente für den subtropischen Typ und in der strömungsnormalen Komponente für den Typ mit vereinigten polaren und subtropischen Zweigen.

Die mittleren Fluktuationen des Windes sind für die strömungsnormale Komponente in folgenden Bereichen der "Quer-Skala" größer:

1. im subtropischen Zweig des Strahlstromes, Kollektiv 6 ;
2. im polaren Zweig des Strahlstromes, Kollektiv 2 ;
3. im antizyklonalen Scherungsbereich des polaren Strahlstromes rechts vom Strom, Kollektiv 3 .

Die mittleren Fluktuationen des Windes sind für die strömungstangentiale Komponente in folgenden Bereichen der "Quer-Skala" größer :

1. im Strahlstrom, sofern subtropischer Zweig und polarer Zweig vereinigt auftreten, Kollektiv 8 ;
2. im antizyklonalen Scherungsbereich des polaren Strahlstromes rechts vom Strom, Kollektiv 3 .

Die atmosphärische " Turbulenz " scheint für bestimmte Gebiete nicht isotrop zu sein. Jedoch sollten die durch erhöhte mittlere Fluktuationen festgestellten " Vorzugsrichtungen " nicht als "anisotrope Turbulenz" gedeutet werden. Das gilt besonders, weil Strahlströme durchflogen wurden. Ein Frontalstrom spaltet sich in die beiden Äste (polar, subtropisch), wenn sich die

Randscherung (Scherungsvorticity) in Krümmungsvorticity verwandelt. Eine nordhemisphärische Zirkulation, in der zwischen Strahlströmen polaren und subtropischen Typs klar unterschieden werden kann, neigt deshalb stärker zu meridionalen Strömungen als eine Zirkulation, in der nur ein Strahlstrom in mittleren Breiten festzustellen ist. Die Krümmungen im polaren und subtropischen jetstream erzeugen höhere Fluktuationen in der strömungsnormalen Komponente während des Durchfluges; dagegen bedingt die stärker zonale Anordnung eines nicht gespaltenen Frontalstromes (Kollektiv 8) höhere Fluktuationen in der strömungstangentialen Komponente während des Durchfluges.

Die mittleren Fluktuationen des Windfeldes sind hoch im antizyklonalen Scherungsbereich des polaren Strahlstromes (Seite 74, Bild 7, Kollektiv 3). Der Störungseffekt eines im geostrophischen Gleichgewicht gestörten Luftmassenelementes ist im antizyklonalen Scherungsbereich wesentlich größer als im zyklonalen. Konvektive Initialstörungen treten dort auf, wo eine hinreichende Labilisierung erzeugt wird. Diese Labilisierung ist über dem Atlantik bis in mittelhohe Schichten am stärksten, wenn kalte Luftmassen aus den Polargebieten über das warme Wasser des Golfstromes fließen. Diese Situation ist für den antizyklonalen Scherungsbereich des polaren jetstream sehr viel häufiger als für den antizyklonalen Scherungsbereich des subtropischen jetstream gegeben.

Dagegen unterscheiden sich die erflogenen Winde von den geostrophischen Winden (der Wetterkarte entnommen) am geringsten in den seitlichen Randgebieten der Strahlströme, besonders in den antizyklonalen Scherungsbereichen. Höhere mittlere Streuungen für die vektoriellen Differenzen zwischen erflogenen und geostrophischen Winden liegen für die Bereiche zwischen zwei Strahlströmen (polar - subtropisch) vor. Sie beruhen hauptsächlich auf Wetterkartenfehlern, genauso wie die höheren "ageostrophischen Winde" in den Strahlstromkernen. Die genaue Lage der Kerne ist über dem Atlantik schwer festzustellen. Die höchsten Streuungswerte weist der Kern des polaren Strahlstromes auf (Einfluß der Initialstörungen auf den Gleichgewichts-jetstream, Raethjen (43)). Die Werte für die übrigen Kerne sind aber nicht viel niedriger, so daß bei einer Vermehrung des Materials unter Berücksichtigung anderer Wetterlagen sehr wohl eine geringe Verschiebung eintreten kann. Eine zusätzliche Unterscheidung der Daten sowohl nach strömungsparalleler als auch nach vertikaler Lage wäre wünschenswert, jedoch liefert hierzu das Material keine statistisch gesicherten Ergebnisse.

3.6.4 Statistische Betrachtung der Fluktuationen und ageostrophischen Komponenten des Windfeldes in der "Parallel-Skala"

Die Aufgliederung der Windwerte in Kollektive, die parallel zur Achse des Strahlstromes liegen, ergibt ungefähr gleiche Werte im zyklonalen und antizyklonalen Strömungsbogen für die mittlere Fluktuation des Windes. Dieses Bild mag sich verschieben, wenn eine dreidimensionale Aufteilung der Werte vorgenommen werden kann (vertikal, normal und tangential zur Strömung). Die Zahl der vorliegenden Messungen reicht hierzu nicht aus. Die Ergebnisse mögen sich auch verändern, wenn statt einer Ordnung nach der Makrostruktur eine Ordnung nach der Mesostruktur vorgenommen wird. Eine derartige Feinanalyse ist heute noch unmöglich.

Die statistische Untersuchung des vorliegenden Materials in den beschriebenen s-förmigen Strömungsbild zeigt keine erhöhten mittleren Fluktuationen des Windes in Trögen. Sicherlich sind in einzelnen Fällen große Windunterschiede in Trögen nachzuweisen (siehe Weber (70)), jedoch muß auch bei antizyklonaler Isohypsenkrümmung mit größeren kleinräumigen Veränderungen im Windfeld gerechnet werden. Große Fluktuationen des Windes treten im antizyklonalen Bogen an Stellen mit kleinen Krümmungsradien der Höhenströmung (nach der Wetterkarte) auf. Da nur wenige derartige Fälle im Material vorliegen, verursachen sie nur wenig höhere Werte der mittleren Fluktuation des Windes. Sie haben statistisch aber mehr Gewicht als die größeren mesostrukturellen Windunterschiede in Troglagen.

Die Abweichungen der erflogenen Winde von den geostrophischen Winden sind bei antizyklonaler Strömung etwas höher als bei zyklonaler Strömung. Die Vergleiche mit dem Gradientwind führen am Ende des antizyklonalen Strömungsbogens zu besseren, allgemein zu nur wenig schlechteren Ergebnissen. Lediglich für Fälle im Anfang des antizyklonalen Bogens liegen die Werte der mittleren Streuung für die Differenzvektoren zwischen Gradientwinden und erflogenen Winden wesentlich über denjenigen für die Differenzvektoren zwischen geostrophischen Winden und erflogenen Winden. Das Gradientwindfeld ist also besonders im ersten Teil des antizyklonalen Bogens ungenau (hier könnte systematische Aufgleitbeschleunigung vorliegen).

4. Die Genauigkeit der Höhenwetterkarte

Die Fehler einer Vorhersagekarte werden wesentlich von den Fehlern in der synoptischen Wetterkarte bestimmt, die zur Prognose benutzt wird. Es ist deshalb notwendig, den ursächlichen Zusammenhang dieser Eingangsfehler zu untersuchen, die sich aus den Meßfehlern der Radiosonden, Übermittlungsfehlern und Fehlern bei der Eintragung der gemessenen Daten in die Wetterkarte sowie deren fehlerhafte Interpretation bei der Analyse der Wetterlage zusammensetzen. Die Meßfehler der Radiosonden wurden anläßlich mehrerer Weltvergleiche von Radiosonden in Payerne (Schweiz) durch Beelitz (3) beschrieben. Einen ausgezeichneten Beitrag zu diesem Problem lieferte Bergren (4) für die 500-mb-Fläche. Aus mehr als 20 unabhängigen Analysen der gleichen Wetterlage durch verschiedene Meteorologen gelang es, den mittleren Fehler des Druckfeldes und des geostrophischen Windfeldes abzuschätzen.

Das durch die Deutsche Lufthansa erflogene Material gestattet, die Wetterkartenfehler über eine Reihe von Wetterlagen zu ermitteln. Dabei kann der mittlere Fehler eines geostrophischen Windes m_g, der aus dem Isohypsen-Abstand in der Wetterkarte errechnet wird, nach dem Fehlerfortpflanzungsgesetz aus den Quadraten der Abweichungen zwischen erflogenem und geostrophischem Wind abzüglich des mittleren Fehlers des erflogenen Windes bestimmt werden:

$$(4.1) \qquad m_g^2 = \sum_i (f_i - g_i)^2 / (n-1) - m_f^2 \; ;$$

wobei
- f_i Einzelwert des erflogenen Windes,
- g_i Einzelwert des geostrophischen Windes,
- n Zahl der Fälle,
- i Index des Einzelwertes.

Dieser einfachen Überlegung stellen sich einige Schwierigkeiten entgegen. Zwar ist der mit dem Flugzeug gemessene Wind zu eichen, jedoch ist das Verfahren sehr umständlich und nur unter großem finanziellem Aufwand möglich. Es muß deshalb auf Grund der mittleren instrumentellen Fehler (Fühler, Anzeige- und Registriergeräte im Flugzeug) nach dem Fehlerfortpflanzungsgesetz der mittlere Fehler für den erflogenen Wind abgeschätzt werden. Diese theoretisch gefundenen Werte wurden durch einen Begleiter, der während mehrerer Flüge zwischen Frankfurt und San Francisco in der Flugzeugkanzel die auf den Fluginstrumenten angezeigten Werte protokollierte, bestätigt. Es gelang auf einem Flug zwischen Montreal und San Francisco am 27.3.1963 - 00 GMT die aufsteigende Radiosonde der aerologischen Station 72.637 (Flint) zu sichten. Die vektoriellen Differenzen zwischen dem Radiosondenwind in 200 mb (Flugniveau) und erflogenem Wind (nächstliegender Wert im berechneten Gitternetz mit Punktfolge von 6 Zeitminuten) beträgt 1,7 m sec^{-1}. Der Betrag des Differenzvektors Radiosondenwind-erflogener Wind (über ungefähr 100 km gemittelt) lautet 0,5 m sec^{-1}. Diese beiden Fehlerangaben zeigen, wie schwierig es ist, gut vergleichbare Daten aus Messungen mit der Radiosonde (vertikale Integration des Windes) und mit dem Flugzeug (horizontale Integration) zu erhalten.

Das Druckfeld einer Höhenwetterkarte wird bei einem Stationsabstand über dem nördlichen Atlantik von mehr als 1000 km nur sehr grob analysiert. Deshalb können die entnommenen Daten des geostrophischen Windes oder des Gradientwindes keine Feinstruktur zeigen und die sich ergebenen Unterschiede zwischen erflogenen und geostrophischen Winden sollten nicht unbedingt als "Fehler der Wetterkarte" interpretiert werden. Eine dichte Folge von Meßpunkten würde das Bild der Zirkulation verfeinern. Es ist notwendig, das rechte Vergleichsmaß zwischen geostrophischen und erflogenen Winden zu suchen.

Die erflogenen Winde werden deshalb über die im Abschnitt 3.3 beschriebenen Glättungsformeln (3.3.1) bis (3.3.3) ausgeglichen beziehungsweise über die gleichen Intervalle nach der Gleichung (3.3.4) gemittelt, und jeweils mit den geostrophischen Winden verglichen. Dieses Verfahren ist scharf von jenen Nebenuntersuchungen zu trennen, die zur Festlegung der bestmöglichen Intervallstrecke für die Bestimmung eines erflogenen Windes, sowie des Abstandes der Gitterpunkte in der Aufbereitung der Registrierstreifen dienten. Wie in Abschnitt 9.2 beschrieben, ergab sich dabei ein zweckmäßiger Abstand der Gitterpunkte von sechs Minuten Flugzeit und eine Mittelung der registrierten Parameter über ein Intervall von ±1 Minute. Die durch Flugmanöver verursachten, störenden Einflüsse werden bei diesem Abstand der Meßpunkte und entsprechender Glättung routinemäßig während der Datenaufbereitung unterdrückt; die Meßpunkte liegen dabei für meteorologische Zwecke dicht genug. Jetzt muß dagegen ein erflogener Wind gesucht werden, der gut in das unabhängig gezeichnete System "Wetterkarte" hineinpaßt.

Falls die mittlere Streuung der vektoriellen Differenzen zwischen erflogenen und geostrophischen Winden hauptsächlich durch die mesostrukturellen Fluktuationen des Windfeldes bestimmt wird, können durch Glättung und Mittelung der erflogenen Daten die Streuungswerte erniedrigt werden. Wenn das benutzte Intervall am besten in die Maschenweite des aerologischen Netzes hineinpaßt, ergibt sich für die mittlere Streuung der Differenzvektoren "geglätteter erflogener Wind-geostrophischer Wind" ein Minimum. Das geostrophische Feld einer Wetterkarte würde in diesem Fall das mittlere Stromfeld gut beschreiben. Die Fehler in der Analyse wären unbedeutend. Die Fluktuationen des Windfeldes würden nur für einzelne Windwerte bemerkenswert sein.

Falls die mittlere Streuung der Differenzvektoren zwischen erflogenen und geostrophischen Winden durch eine Glättung der erflogenen Daten nicht merklich reduziert werden kann, wird der hohe Streuungswert entweder durch falsche Flugzeugwinde oder durch falsche (ungenaue) Analysen der Wetterkarten verursacht. Die erflogenen Winde stimmen mit Radiosonden-Winden gut überein, sofern der räumliche und zeitliche Abstand klein ist. Die Beiträge zu hohen Streuungswerten der Differenzvektoren "geglätteter erflogener Wind - geostrophischer Wind" würden durch Analysenfehler geliefert.

Es wäre deshalb anzunehmen, daß die ungenaue Zeichnung der Isopotentialen (falsche Krümmung, falscher Abstand), sowie unberücksichtigte Verlagerungen des Stromfeldes große Fehler in das Kollektiv hineintragen.

Zusammenfassend seien folgende Beiträge zur mittleren Streuung der vektoriellen Differenzen zwischen erflogenen und geostrophischen Winden erwähnt:

1. Fehler bei der Radiosonden-Messung,
2. Fehler bei der Übermittlung und Eintragung der Daten,
3. Fehler bei der Analyse der Wetterkarte,
4. Fehler durch große Maschenweite des aerologischen Netzes, Fluktuationen des Windes,
5. Fehler durch Druckfeldwanderungen und Verlagerungen des Stromfeldes, zeitlicher Abstand der Flugzeugmessung vom Haupttermin (Aufstieg der Radiosonde),
6. Verschiedenheit der Meßverfahren, (horizontale Integration bei erflogenen Winden, vertikale Integration bei Radiosonden-Winden),
7. Fehler bei der Windmessung mit dem Flugzeug,
8. Adataption, ageostrophische Winde, sonstige meteorologische Erscheinungen.

Die Beiträge 1 bis 3 sind durch Berggren (4) beschrieben. Der Beitrag 4 ist durch die Größe der mittleren Fluktuationen abschätzbar. Der Beitrag 5 kann durch Ordnung der Daten nach ihrem zeitlichen Abstand von einem synoptischen Haupttermin erfaßt werden. Der Beitrag 6 wird aus dem Vergleich der erflogenen Winde mit Radiosonden-Winden (direkter Uberflug der Station während des Aufstieges) gefunden; er ist kleiner als $3\,m\,sec^{-1}$, wahrscheinlich liegt er im Mittel unter $1\,m\,sec^{-1}$. Der Beitrag 7 wird in Abschnitt 9 ausführlich behandelt; er ist kleiner als $2{,}9\,m\,sec^{-1}$.

Es ist also unabhängig von dem Verfahren nach Berggren möglich, die Beiträge 1 bis 3 zum mittleren Fehler der Wetterkarte aus den erflogenen Daten abzuschätzen, sofern die unter Punkt 8 genannten Beiträge unbedeutend sind.

Eine erste Information über die Güte der Höhenwetterkarte gewähren Korrelations- und Regressionsrechnungen. Der geostrophische Wind und der Gradientwind wurden skalar und vektoriell mit dem erflogenen Wind verglichen. Die Korrelationsfaktoren liegen zwischen 0,64 und 0,80 (Zufallshöchstwert kleiner als 0,095 bei 1000 Werten; in allen Kollektiven sind mehr als 2000 einwandfreie Windvergleiche vorhanden). Der skalare Vergleich ergibt erwartungsgemäß etwas schlechtere Ergebnisse. Die Korrelationsfaktoren sind für die nördliche Komponente günstiger als für die östliche Komponente. Die Korrelation des erflogenen Windes ist zum geostrophischen Wind besser als zum Gradientwind. Die höchsten Korrelationsfaktoren werden für eine Glättung der Daten über eine Flugstrecke von ungefähr 600 km erzielt, wie Tabelle 2 zeigt.

Tabelle 2: Korrelationsfaktoren erflogener Wind - geostrophischer Wind und Gradientwind

			vektoriell	
	Glättung	skalar	Nord	Ost
erflogener Wind - geostrophischer Wind	-	0,684	0,797	0,738
	12 min		0,807	0,746
	48 min		0,820	0,759
	84 min		0,814	0,752
erflogener Wind - Gradientwind	-	0,644	0,764	0,714
	12 min		0,773	0,722
	48 min		0,784	0,732
	84 min		0,779	0,728

Anmerkung: In Spalte "Glättung" ist die Flugzeit in Minuten eingetragen, über welche die Glättung vorgenommen wurde.

Bei Annahme verhältnismäßig hoher Fehler der Fluginstrumente ergibt sich ein mittlerer Fehler des erflogenen Windes, der kleiner als 3 m sec^{-1} ist. Dieser geringe mittlere Fehler der Einzelmessung ist zur Abschatzung des mittleren Fehlers der Wetterkarte zunächst vernachlassigbar. Da der "wahre Wind" unbekannt ist, soll der erflogene Wind als "wahrer Wert" gelten, das heißt er soll nicht mit Fehlern behaftet sein. Dieses Verfahren liefert einen zu großen Wert für den mittleren Fehler des geostrophischen Windes (Gradientwind) aus der Wetterkarte. Wären die Fehler des geostrophischen Windes aus der Wetterkarte und die des erflogenen Windes gleich groß, nur von entgegengesetzten Vorzeichen, läge also der "wahre Wert" genau in der Mitte beider Daten, müßte mit der halben Differenz zwischen beiden Daten (erflogener bzw. geostrophischer Wind) in die Fehlerrechnung eingegangen werden. Es erfolgt bewußt die Abschätzung eines größtmoglichen mittleren Fehlers. Am Ende der Rechnung wird nach Gleichung (4.1) der Streuungswert reduziert, um den mittleren Fehler des geostrophischen Windes zu erhalten.

Aus den Quadraten der vektoriellen Differenzen zwischen erflogenen und geostrophischen Winden ergibt sich die mittlere Streuung von 15,5 m sec^{-1} ; zwischen erflogenen Winden und Gradientwinden diejenige von 17,0 m sec^{-1} .

Die Verlagerung der Druckgebilde ist in allen Rechnungen für die vorgelegte Arbeit unberücksichtigt geblieben. Lediglich wurde, wie schon erwähnt, bei einem zeitlichen Abstand von mehr als 5 Stunden das Mittel aus beiden zur Verfugung stehenden Analysen verwertet. Diese allgemeine Vernachlässigung ist insofern gerechtfertigt, als die Werte der mittleren Streuung bei einem zeitlichen Abstand eines Meßpunktes von mehr als einer Stunde zu einem synoptischen Haupttermin über 14 m sec^{-1} liegen und um höchstens 2 m sec^{-1} verbessert werden können.

Die mittlere Streuung der vektoriellen Differenzen erflogener - geostrophischer Wind beträgt für Werte in einem zeitlichen Abstand von weniger als einer Stunde zu einem synoptischen Haupttermin 10,9 m sec^{-1} . Sie steigt bei größerem zeitlichen Abstand bis 16,6 m sec^{-1} . Durch Mittelung der Daten aus vor- bzw. zurückliegender Wetterkarte sinkt der Wert für die mittlere Streuung auf 14,6 m sec^{-1} (zeitlicher Abstand vom synoptischen Haupttermin 5 bis 6 Stunden).

Tabelle 3: Mittlere Streuung der vektoriellen Differenzen zwischen geostrophischen und erflogenen Winden im zeitlichen Abstand von synoptischen Hauptterminen (keine Glättung)

Zeitlicher Abstand in Stunden	0 - 1	1 - 2	2 - 3	3 - 4	4 - 5	5 - 6
Mittlere Streuung in m sec^{-1}	10,9	14,5	16,2	15,8	16,6	14,6

Die in Tabelle 3 angegebenen mittleren Streuungswerte sind bei dem geringen mittleren Fehler des erflogenen Windes (kleiner 3 m sec^{-1}) um weniger als 1 m sec^{-1} zu reduzieren, um die mittleren Fehler des geostrophischen Windes aus der Wetterkarte zu erhalten. Dieser mittlere Fehler der Einzelmessung setzt sich hauptsächlich aus Analysenfehlern in der Feinstruktur, falsche Lage der Zonen mit hohen Windgeschwindigkeiten, weniger aus groben Analysenfehlern und Meßfehlern der Radiosonden zusammen. Die Berücksichtigung der Druckfeldverlagerungen innerhalb zwölf Stunden verbessert die geostrophischen Daten nur wenig, weil die genaue Lage der Starkwindfelder schon in der Ausgangskarte schwer zu erfassen ist.

Die mittlere Streuung der Differenzvektoren zwischen geostrophischen Winden und erflogenen Winden, welche aus der Abdrift des Flugzeuges navigatorisch bestimmt werden, beträgt bei einem zeitlichen Abstand von weniger als zwei Stunden für das 500-mb-Niveau ungefähr 11 m sec^{-1} . Dieser Wert wurde aus Navigationsprotokollen von Flügen der Deutschen Lufthansa mit dem Flugzeugtyp L 1049 G (Superconstellation) ermittelt (13). Die Flüge fanden in den Monaten Oktober 1959 bis März 1960 über dem nördlichen Atlantik statt. Die Höhe dieses Wertes wird wesentlich durch die Ungenauigkeit in der Navigation bestimmt. Nach den Ergebnissen war ein mittlerer Fehler des geostrophischen Windes aus der Wetterkarte von ungefähr 9 m sec^{-1} zu erwarten. Es konnten damals keine Ursachen für diesen hohen Fehlerwert der Einzelmessung angegeben werden. Die Ergebnisse gelten für die 500-mb-Fläche.

Die sehr viel genaueren mit dem Doppler-Radar-Gerät erflogenen Winde ergeben zwar ein feineres Windfeld in Tropopausenhöhe, die Übereinstimmung mit der Wetterkarte ist aber nicht verbessert. Eine geringe Verbesserung gelingt, sofern die Daten nach den im Abschnitt 3.3 (Seite 53) angegebenen Glättungsformeln übergreifend über ein Intervall von 48 Minuten Flugzeit (ungefähr 600 km) gemittelt werden.

Tabelle 4: Mittlere Streuung der vektoriellen Differenzen zwischen geostrophischen und erflogenen Winden bei Anwendung übergreifender Mittelung oder Glättung der Daten

Intervall in min	1	12	24	48	84
Mittlere Streuung bei Glättung	15,4	15,0	-	14,3	14,8 m sec^{-1}
Mittlere Streuung bei Mittelung	15,4	15,3	15,6	16,6	- m sec^{-1}

Wie Tabelle 4 zeigt, sind die erflogenen Daten für die heute gebräuchlichen Wetterkarten besser geeignet, falls sie nach Gleichung (3.3.2) geglättet und nicht gemittelt werden. Diese Distanz entspricht etwa der Maschenweite des aerologischen Netzes. Die Zahl der erflogenen Winde mit einem zeitlichen Abstand von weniger als einer Stunde zu einem synoptischen Haupttermin ist zu gering, so daß zur Berechnung der Tabelle 4 alle verfügbaren Werte verwendet werden mußten. Die Streuungswerte dürften durch übergreifende Glättung auch für zeitlich besser übereinstimmende Daten in gleicher Größenordnung verringert werden (1 bis 2 m sec^{-1}).

Die Güte der Wetterkarte kann durch ein dichteres Beobachtungsnetz erheblich verbessert werden. Die Ergebnisse des Abschnitts 3 zeigen größere Ungenauigkeiten für östliche Strömungen. Dagegen sind für südliche Strömungen geringere mittlere Streuungswerte festzustellen. Diese geringeren Werte müßten bei besserem und zahlreicherem Datenangebot sowie feinerer Analyse für alle Windrichtungen zu erreichen sein. Die mittlere Streuung der vektoriellen Differenzen zwischen geostrophischen und erflogenen Winden beträgt in diesem Fall 9 m sec^{-1}. Nach dem Fehlerfortpflanzungsgesetz können von diesem Wert ein Fehlerbeitrag, der durch die grobe Maschenweite des aerologischen Beobachtungsnetzes entsteht (1 bis 2 m sec^{-1}), ein Fehlerbeitrag, der durch die Zeitdifferenz zwischen Messung mit der Radiosonde und mit dem Flugzeug entsteht (2 bis 3 m sec^{-1}) und der mittlere Fehler des erflogenen Windes (3 m sec^{-1}) abgezogen werden. Der sich ergebende Wert von ungefähr 8 m sec^{-1} stellt den mittleren Fehler des geostrophischen Windes dar (für Einzelwerte aus der Wetterkarte).

Die mesostrukturellen Fluktuationen des Windes lassen eine bessere Übereinstimmung zwischen einem erflogenen und einem geostrophischen Wind nicht zu (mittlere Fluktuation des Windes siehe Abschnitt 3). Der mittlere Fehler eines geostrophischen Wetterkarten-Windes liegt zwischen den Grenzen 4 m sec^{-1} und 8 m sec^{-1}, sofern erflogene Winde genutzt werden und der Abstand der Meßpunkte längs der Flugwege nicht größer als 100 km ist. Diese Forderungen sind nur durch automatische Abrufverfahren, internationalen Austausch , moderne Datenverarbeitungsanlagen und Einsatz von Großrechenanlagen zu verwirklichen.

Heute muß für den einzelnen geostrophischen Wind, wie er der Wetterkarte entnommen wird, ein mittlerer Fehler von 10 m sec^{-1} angesetzt werden. Diese Werte erhöhen sich bei größerem zeitlichen Abstand zwischen Messung und synoptischem Haupttermin auf 15 m sec^{-1}. Der mittlere Fehler des geostrophischen Windes aus der Wetterkarte ist besonders groß in der Nähe der Tropopause und im Bereich der Strahlströme.

Bild 4 (Seite 65) zeigt die mittleren Streuungen der vektoriellen Differenzen zwischen geostrophischen und erflogenen Winden in Abhängigkeit von der Geschwindigkeit des erflogenen Windes (rechts fett ausgezogen). Die Werte der mittleren Streuungen nehmen für Geschwindigkeiten über 30 m sec^{-1} stark zu. Derartige hohe Windgeschwindigkeiten treten in Strahlströmen auf. Abschnitt 3.6.3 bestätigt besonders hohe mittlere Streuungen der vektoriellen Differenzen zwischen geostrophischen und erflogenen Winden für die Bereiche der Strahlströme. Der mittlere Fehler eines geostrophischen Windes nach der Wetterkarte liegt nach diesen Ergebnissen im Bereich der Strahlströme über 20 m sec^{-1} und ist für die Komponente tangential zur Strömung um ungefähr 8 m sec^{-1} höher als für die Komponente normal zur Strömung.

Diese hohen mittleren Streuungswerte haben nur für Einzelwerte große Bedeutung. Die statistisch ermittelten Ergebnisse sagen nicht, daß die Wetterkarte in ihrer Gesamtheit derartig hohe Fehler aufweist. Die Makrostruktur der Atmosphäre ist meist richtig erfaßt, nur die Lage der Tiefdruckzentren kleinerer Ausdehnung und die Erstreckung der Starkwindzonen ist häufig falsch. Der Druckgradient zwischen den aerologischen Aufstiegen wird gleichmäßig über die Distanz verteilt. Der Abstand der aerologischen Stationen ist über dem Atlantik zu groß. Es gelingt deshalb in allen Fällen, analysierte Wetterkarten umzuzeichnen und die erflogenen Winde genau einzupassen. Häufig genügt eine geringe zeichnerische Verlagerung des jetstream oder eine Änderung der Isohypsenkrümmung, um die erflogenen Winde den geostrophischen Winden anzupassen. Komplizierte Strömungsfelder (kleinräumige Wirbel) sind kaum zu entwerfen.

Das Gradientwindfeld liefert höhere mittlere Streuungswerte (um 1 bis 2 m sec^{-1} höher) verglichen mit erflogenen Winden. Der Krümmungsradius ist in einer Höhenwetterkarte schwer richtig zu erfassen, aber sehr leicht entsprechend den erflogenen Daten zu verändern (geringe Verlagerung des Strömungsbildes, Mäander). Die erflogenen Winde können unter etwas größerem Aufwand auch dem Gradientwindfeld sehr gut eingepaßt werden.

5. Diskontinuitäten im Höhenwindfeld

In jüngster Zeit werden von einigen Instituten Isohypsen mit scharfen Knickstellen besonders im Bereich der Höhentröge in die Höhenwetterkarte gezeichnet. Dabei wird das Verfahren, die Isobaren längs der Frontlinien zu knicken, auch für Topographien der Druckflächen über 500 mb verwandt. Diese für den Beschauer einprägsame zeichnerische Lösung der meteorologischen Verhältnisse in der höheren Atmosphäre kann nicht kritiklos hingenommen werden. Der Windsprung am Boden beim Durchgang einer Front wird hauptsächlich durch starke vertikale Umlagerungen in den unteren Luftschichten erzeugt. In der unteren Stratosphäre sind derartige vertikale Transporte wegen der meist stabilen Schichtung undenkbar. Alle Überlegungen müssen berücksichtigen, daß die schnell fließenden Luftteilchen träge sind, ihren Kurs also nicht sprunghaft ändern können. Die Kontinuitätsgleichung muß stets (dreidimensional) erfüllt sein.

Durch die Trägheit der Massenteilchen würde innerhalb kurzer Zeit eine scharfe "stationäre Knickstelle" in einer horizontal fließenden Luftströmung geglättet werden, wenn nicht spontan alle heranfließende Masse entweder horizontal oder vertikal fortgeschafft würde. Deshalb sind die durch Isohypsenknicke dargestellten Höhenscherungsflächen (wie Bodenfronten) im Windfeld nicht stationär, sondern die Scherungsflächen bewegen sich mit der flächensenkrechten Windkomponente. Die Krümmungsradien der Strombahnen können bei den geringen Windgeschwindigkeiten in der Trogachse wenige Kilometer betragen, so daß die herrschende Strömung in der Natur zwar nicht geknickt sondern "scharf gebogen" ist, aber in einer üblichen Wetterkarte "geknickt" dargestellt werden darf. Die Zeichnung scharfer Knicke der Isohypsen in den Höhenwetterkarten kann also aus beiden genannten Gründen (1. Scherungsflächen sind instationär, 2. Maßstab der Wetterkarte) in Einklang mit der Physik stehen. Die Zahl und die Genauigkeit der Beobachtungen werden aber in den meisten Fällen nicht ausreichen, um derartige Scherungsflächen zu analysieren.

Das Verhalten eines Flugzeuges dürfte beim Durchflug einer Scherungsfläche mit starken Windunterschieden zwischen der Vorder- und Rückseite seinem Flugverhalten an einer Strömungsdiskontinuität ähneln: Die Auftriebskraft ändert sich sprunghaft. Im Bereich schmaler Höhentröge treten starke Erscheinungen der "Clear Air Turbulence" auf, die auf horizontale und vertikale Beschleunigungen der Luftquanten deuten und die sich durch Erschütterungen am Meßträger kundtun. Die Erschütterungen sind die Folge veränderter Auftriebskräfte und brauchen nicht durch Beschleunigungen der umgebenden Luft verursacht zu sein. Sie entstehen auch beim Durchfliegen von Luftmassen verschiedener Windgeschwindigkeiten. Im letzten Fall ändert sich kurzzeitig die Auftriebskraft, weil die horizontale Geschwindigkeit des Flugzeuges zunächst konstant bleibt, aber sowohl der Anstellwinkel (Winkel, den die Tragfläche mit der Bewegungsrichtung der anströmenden Luft bildet), als auch die Flugzeugeigengeschwindigkeit (Relativgeschwindigkeit des Flugzeuges zur Luft) sich ändern. Falls der Auftrieb ungleich dem Gewicht des Flugzeuges ist, folgen vertikale Beschleunigungen des Flugkörpers. Horizontale Böen verursachen zwar geringere vertikale Beschleunigungen des Flugzeuges als vertikale Böen, jedoch sollten besonders wegen des bevorstehenden Überschallflugverkehrs die Wirkungen horizontaler Windunterschiede nicht unbeachtet bleiben.

Isobare Windunterschiede können an Sperrschichten (z.B. Tropopause) herrschen, die horizontal oder isentrop liegen. Die verschieden gerichteten Strömungen der Luft über und unter der Sperrschicht dürften die Neigung der Grenzfläche einnehmen. Obwohl der Meßträger im isobar gesteuerten Flug einen scharfen Windsprung während des Schnittes der Sperrschicht registriert, weisen die getrennten Strömungen oberhalb und unterhalb der Sperrschicht keine "Diskontinuitäten" in einer strömungsparallelen Fläche auf. In einer Isobarenfläche wäre es gestattet, die Schnittstellen der Fläche mit Sperrschichten oder Grenzflächen verschieden strömender Luftmassen durch Knicke in den Isohypsen zu markieren. Dabei muß man sich klar sein, daß an diesen Stellen (im stationären Fall) der Strömungsvektor die dargestellte Isobaren-Fläche schneidet, oder sich die Flächen mit der flächensenkrechten Windkomponente bewegen.

Es sei auf die Darstellungen von Höhenwetterkarten mit Windscherungsflächen in der freien Atmosphäre durch Sußebach (65) hingewiesen. Sußebach benutzte unveröffentlichte Manuskripte des Verfassers und Flugzeug-Wind-Registrierungen, die auch in dieser vorliegenden Arbeit verwendet werden.

Windscherungen, welche durch Flugzeuge registriert werden, bedürfen einer harten Kritik. Die Flugzeugbesatzungen entschließen sich aus flugbetrieblichen Gründen bei Erreichen einer für den Flug großräumigen Wind - Veränderung, die Flughöhe zu wechseln. Die Windbestimmung für den Steig- bzw. Sinkflug und auch für eine nachfolgende Strecke (bis 3 Minuten) ist unmöglich, weil die Einstellzeiten für die Flugzeugeigengeschwindigkeit und für die Grundgeschwindigkeit nicht gleich sind und durch die Neigung des Flugzeuges Geschwindigkeitsveränderungen vorgetäuscht werden. Das Anzeige-Gerät zum Doppler-Radar benötigt mehr als zwei Minuten um eine Veränderung der Grundgeschwindigkeit von $50\ m\ sec^{-1}$ anzuzeigen (Genauigkeit der Windbestimmung in Abschnitt 9). Die hohe Einstellzeit des Anzeige-Gerätes zum Doppler-Radar bewirkt eine langsame Veränderung des "errechneten Windvektors", der während Flugmanövern und plötzlichen Geschwindigkeitsänderungen schlecht mit dem tatsächlichen Windvektor (auch bei kurzen Strecken mit Windstille in einem Trog) übereinstimmt. Es kann sogar bei Gegenwind durch einen Steigflug eine starke Windscherung vorgetäuscht werden. Diese Fehlermöglichkeiten blieben in der oben zitierten Arbeit von Sußebach (65) unberücksichtigt.

Die Frage nach eventuellen Diskontinuitäten in der freien Atmosphäre soll mit je einem Beispiel aus dem Makro-, Meso- und Mikro-Bereich erläutert werden. Die Ergebnisse aus Flugzeug-Registrierungen werden anschließend mitgeteilt.

Beispiel 1 (Mikro-Bereich):

Interessante Aufschlüsse über mikrostrukturelle Eigenschaften der Atmosphäre lieferten Rauchabsonderungen aus Raketen, die zu gleicher Zeit in einem seitlichen Abstand von etwa 3 km abgeschossen wurden (35). Die Windprofile sind für alle drei Aufstiege ungefähr gleich. Zwischen den Höhen 31 000 ft und 37 000 ft ergibt sich die Streuung von 0,4 m sec^{-1}; zwischen den Höhen 23 000 ft bis 31 000 ft eine größere von 0,6 m sec^{-1}. Diese Werte einer Einzelmessung sollten für mesostrukturelle Betrachtungen nicht extrapoliert werden.

Beispiel 2 (Meso-Bereich):

Eine Diskontinuität im Windfeld wurde (nach Flohn (22)) über Schleswig am 7.9.1956 mit Radiosonden erfaßt, die im Abstand von etwa 15 Minuten während des Durchganges eines Höhentroges gestartet wurden. Der Wind drehte dabei in der oberen Troposphäre innerhalb 30 bis 40 Minuten (das entspricht bei der herrschenden Windgeschwindigkeit einer horizontalen Distanz von 30 km) um 120°. Nach dieser Winddrehung schwankt die eingetretene Westströmung bis 10 m sec^{-1}.

Beispiel 3 (Makro-Bereich):

Im Bereich schmaler Höhentröge können entgegengerichtete Strahlströme mit Scherungen um 20 m sec^{-1} auf 100 km Entfernung in der Wetterkarte analysiert werden. Ein extremes Beispiel mit hohen Windscherungen über weiten Gebieten West-Europas sowohl in der Vertikalen als auch in der Horizontalen lieferte die Lage vom 28. bis 29.10.1959. Während die englischen aerologischen Stationen nordwest- bis nördliche Höhenwinde über 50 m sec^{-1} meldeten, herrschte über dem Kontinent Südwind, dessen Geschwindigkeiten nur wenig geringer waren. Über Paris bestand am 29.10.1959 um 00 GMT zwischen der 500-mb- und der 400-mb-Fläche eine vertikale Scherung von 180° und ungefähr 50 m sec^{-1} (Tanck (66)).

Das Radiosondennetz vermag aber auch bei dieser extremen Lage keine genauen Aufschlüsse über die Mesostruktur der Strömung zu geben. Es werden lediglich starke Mäander der Makrostruktur erfaßt. Zwar können definitionsgemäß dem Mesobereich alle Erscheinungen zugeordnet werden, die zwischen 10 und 300 km horizontale Ausdehnung fallen, jedoch sollte bei allen Vorgängen in strömenden Medien bedacht werden, daß die Prozesse besonders im Makro- und Meso-Bereich nicht horizontal isotrop sein müssen.

Aus physikalischen Überlegungen sollte nicht starr an den drei Bereichen festgehalten werden. Sie wurden durch die Meßtechnik und nicht wegen meteorologischer Vorgänge bestimmt. Es ist kaum zu erwarten, daß die durch flugbetriebliche Erfahrungen mit Propeller- und Düsenflugzeugen getroffenen Einteilungen für die mit dreifacher Schallgeschwindigkeit fliegende zukünftige Flugzeuggeneration überhaupt sinnvoll bleiben. Die heute als "mesostrukturelle" Erscheinungen bezeichneten Formen des Strömungsbildes werden für das Überschallflugzeug Wirkungen haben, wie sie heute unter dem Sammelbegriff "Clear Air Turbulence" bekannt sind.

Die beschriebenen Fluktuationen des Windes erzeugen für den Überschallflug unangenehme Flugstrecken, die nicht identisch sind mit jenen, die im Propeller- oder Düsenflug festgestellt werden. Sehr viel unangenehmere Folgen für den Überschallflug hätten Diskontinuitätsstellen im Windfeld, weil dort nicht nur der Auftrieb sondern auch die Temperatur an der Außenhaut des Flugzeuges verändert würde. Entscheidend für länger andauernde Flugunruhe (CAT) ist nicht eine spontane Änderung des Windes, sondern ein Wechsel von horizontalen Böen. Wie Raethjen (44) zeigte, haben Horizontalböen im Schnellflug weniger Effekt als Vertikalböen. In der hohen Stratosphäre werden horizontale Böen ein Überschallflugzeug häufiger und stärker als vertikale Böen treffen.

Tabelle 5: Prozentuale Häufigkeitsverteilung der Fluktuationen des Windes

Fehlergrenze	Gauß	Vektorielle Differenzen		
		Betrag	tangential	normal
Mittlere Streuung $m\,sec^{-1}$	-	6,4	4,7	4,4
± m/2	38,0%	38,3%	50,7%	51,5%
± m	68,0%	73,8%	75,3%	75,9%
± 2m	95,0%	96,7%	96,1%	95,6%
± 3m	99,7%	99,5%	99,1%	99,3%

m = Werte der mittleren Streuung (siehe obere Zeile)

Die Tabelle 5 stellt die Häufigkeiten der Windunterschiede zusammen, welche in den Jahren 1962 und 1963 während der Meßperioden für eine Distanz von sechs Minuten Flugzeit, ungefähr 90 km Weg, registriert wurden. Es sind mit Hilfe einer elektronischen Rechenmaschine alle durch Flugmanöver vorgetäuschten Windunterschiede eleminiert und Schreibfehler berichtigt.

Bei Gültigkeit des Gaußschen Fehlergesetzes liegen innerhalb der Grenzen des dreifachen mittleren Fehlers 99,7% aller Resultate. Die mittlere Streuung für die vektorielle Differenz zweier benachbarter erflogener Winddaten beträgt 6,4 $m\,sec^{-1}$. In Spalte 3 der Tabelle 5 ist die Verteilung für den Betrag der vektoriellen Differenzen angegeben. Das Gaußsche Fehlergesetz wird durch das Kollektiv erfüllt.

Große Windscherungen oder Windfluktuationen waren stets mit stärkerer Turbulenz (Flugunruhe, vertikale Beschleunigungen) verknüpft. Größere Schwankungen der Vertikalbeschleunigungsanzeige bedeuten aber nicht gleichzeitig große Fluktuationen oder Anderungen im Meso-Bereich des Windfeldes. Die Empfindlichkeit und die Einstellzeit für die Flugzeugeigengeschwindigkeit (Staudruck) und für die Doppler - Radar - Daten, wie Grundgeschwindigkeit und Driftwinkel, waren in der installierten Apparatur nicht gleich. Deshalb mußte nachträglich eine visuelle Ausgleichung der Staudruckregistrierung bei

Auftreten von "Clear Air Turbulence" vorgenommen werden. Einflüsse starker Bewegungen des Meßträgers um seine Rumpf- oder Tragflächenachse sowie in der Vertikalen werden ausführlicher in Abschnitt 9 behandelt.

Die wenigen Fälle mit sehr großen vektoriellen Differenzen des Windes auf ungefähr 90 km Distanz (6 Minuten Flugzeit) sind keine "Meßfehler in turbulenter Strömung". Bei Anzeigen starker vertikaler Beschleunigungen sind häufig keine großen horizontalen Windunterschiede festzustellen.

Für die Registrierperiode 1962 sind drei Flüge und für die Registrierperiode 1963 achtzehn Flüge mit Gradienten des Windfeldes über $18\,\mathrm{m\,sec^{-1}}$ (auf 6 Minuten Flugzeit) zu verzeichnen. Die Fälle aus dem Jahre 1963 sind in zwölf Höhenwetterkarten des 300-mb-Niveaus zusammengestellt (Bilder 9 und 10 auf den Seiten 96 und 97). In das Zentrum einer Karte ist zur besseren Darstellung ähnlicher meteorologischer Vorgänge die Stelle gelegt, an welcher der große Windunterschied gemessen ist. Ferner gibt die Darstellung zur geographischen Orientierung den nullten und neunzigsten Längengrad sowie den vierzigsten und sechzigsten Breitengrad als angenäherte Grenzen des Operationsgebietes wieder.

Die festgestellten hohen Windscherungen, zusammengestellt in Tabelle 6, liegen in der Randzone des Strahlstromes und in der Nähe eines alten Höhentiefs, meist in der Nähe einer Allschichten-Zyklone, deren vertikale Achse geneigt ist. Dieses Höhentief schafft in seiner Umgebung eine windschwache Zone, während die Strahlströmung in fast unverminderter Stärke vorbeifließt. Hierbei scheint in einigen Fällen (z.B. 8.6.1963) der Strahlstrom fingerförmig in das Gebiet mit schwachen Luftdruckgradienten hineinzuragen. Die Frage, inwieweit hier von der Atmosphäre ein "Freistrahl" produziert wird, vermag mit den Meßergebnissen nicht beantwortet zu werden.

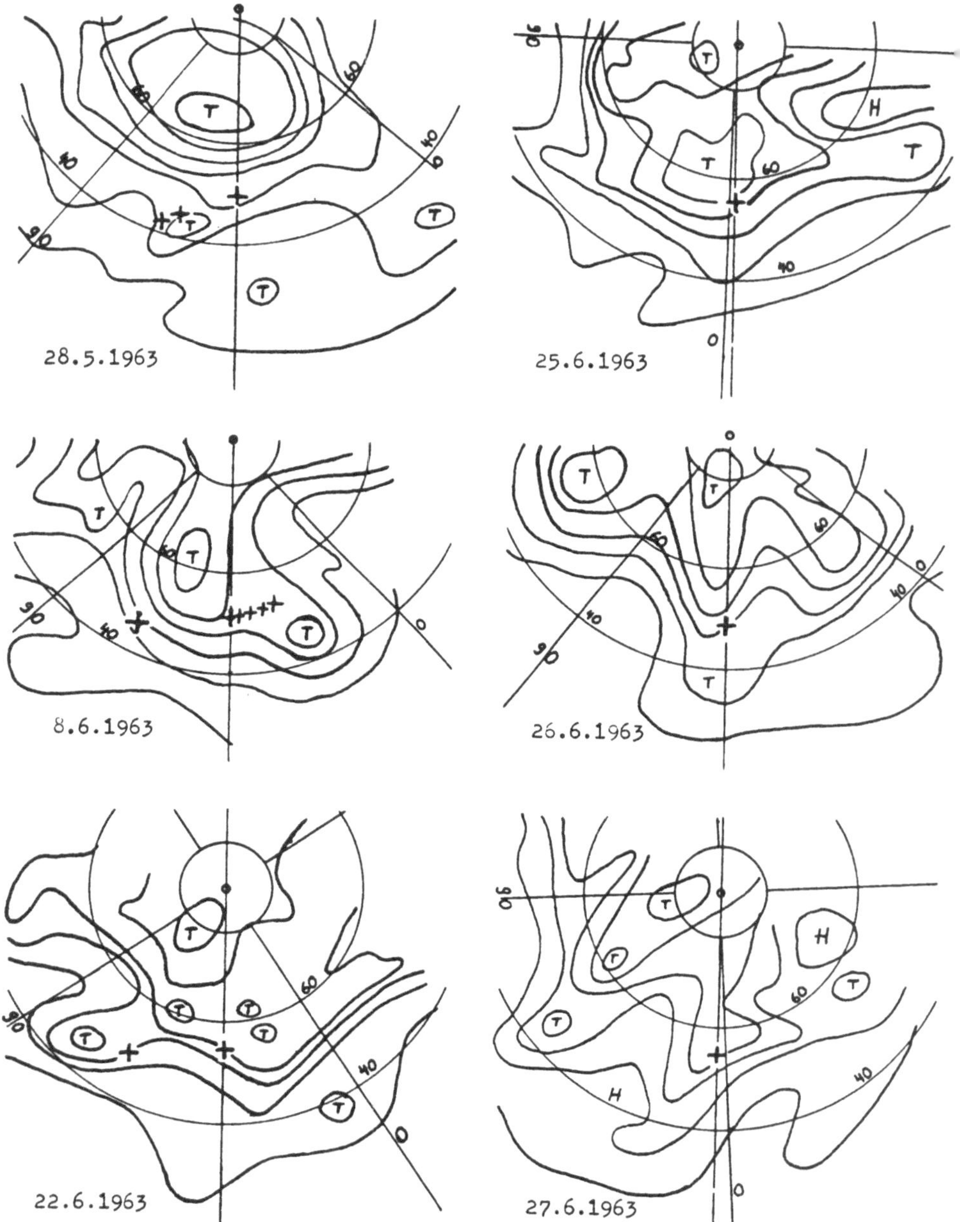

Bild 9: Wetterkarten 300 mb , Lagen mit großen Windunterschieden auf 6 Minuten Flugzeit

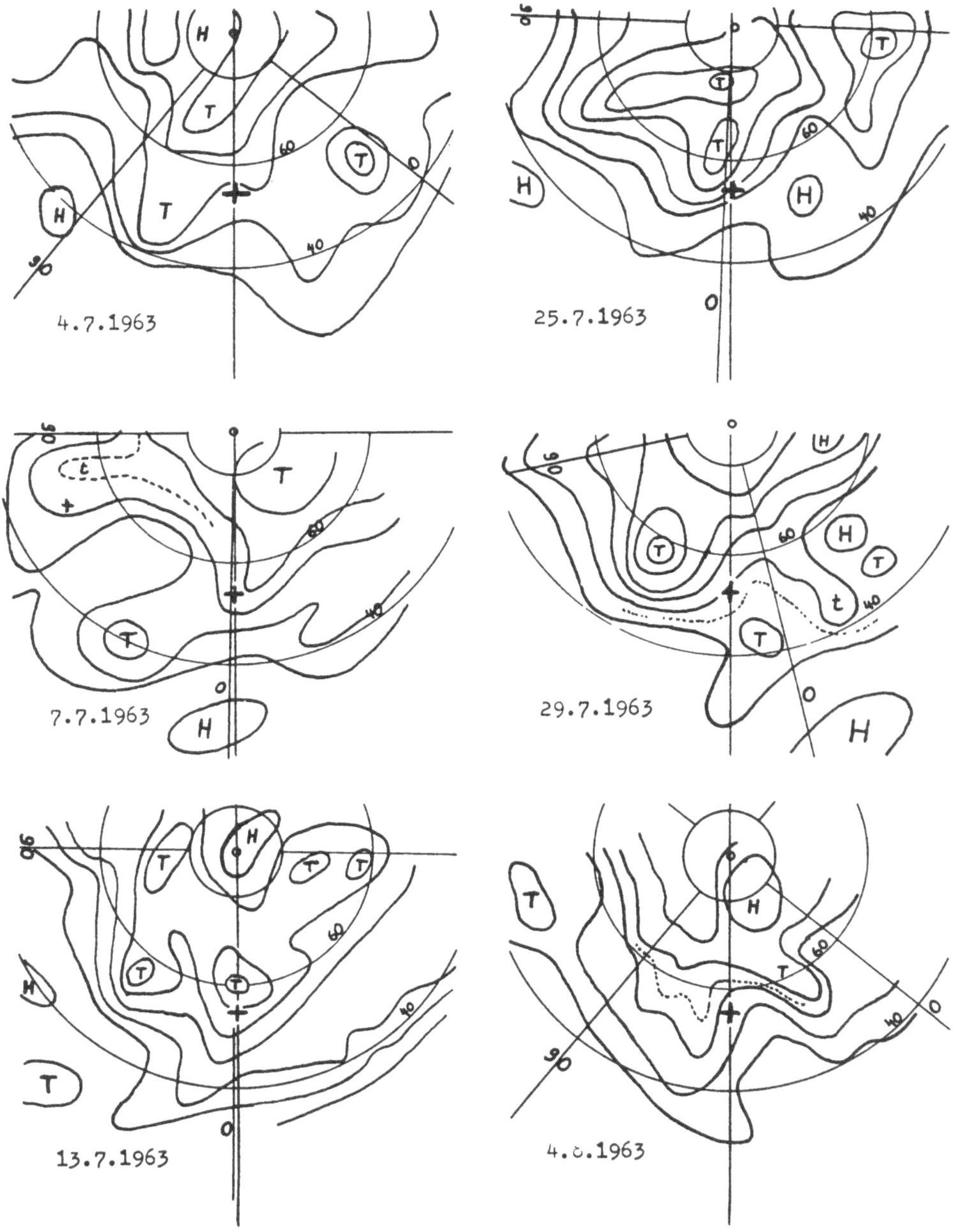

Bild 10: Wetterkarten 300 mb , Lagen mit großen Windunterschieden auf 6 Minuten Flugzeit

Tabelle 6: Zusammenstellung der Windunterschiede über 18 m sec^{-1} (Betrag der vektoriellen Differenz auf 6 Minuten Flugzeit)

Datum	Zeit	Ort	Wind-Differenz (m sec^{-1})	Vertikale Beschleunigung Anzeige Gerät 1	Gerät 2	m sec^{-2}
14.4.1962	3.12	49N-44W	25	0,9	-	0,70
17.4.1962	3.12	51N-31W	18	0,8	-	0,62
10.5.1962	3.30	50N-36W	22	0,4	-	0,31
27.5.1963	18.36	46N-53W	23	-	0,5	0,19
28.5.1963	1.24	44N-68W	21	-	0,5	0,19
	1.42	43N-66W	28	-	0,5	0,19
7.6.1963	19.12	47N-68W	21	-	0,4	0,15
8.6.1963	5.00	51N-42W	37	-	3,0	1,14
	5.06	51N-41W	51	-	3,0	1,14
	5.30	52N-35W	23	-	2,5	0,95
	5.54	52N-30W	22	-	3,0	1,14
	6.06	52N-28W	28	-	2,0	0,76
21.6.1963	17.06	55N-35W	21	-	0,8	0,30
	18.30	51N-49W	19	-	0,8	0,30
22.6.1963	1.24	46N-65W	28	-	2,5	0,95
	1.30	46N-64W	23	-	3,0	1,14
25.6.1963	5.48	49N-01E	21	-	2,8	1,06
	5.54	49N-03E	24	-	0,4	0,15
26.6.1963	3.00	49N-54W	25	-	4,0	1,52
27.6.1963	14.00	54N-02W	30	-	0,5	0,19
29.6.1963	19.18	59N-06W	23	-	0,6	0,23
2.7.1963	14.06	54N-01E	18	-	0,8	0,30
3.7.1963	17.48	55N-53W	18	-	1,4	0,53
7.7.1963	17.00	54N-00E	23	-	0,6	0,23
	23.06	47N-79W	18	-	2,5	0,95
11.7.1963	21.00	46N-63W	19	-	1,2	0,46
13.7.1963	13.48	54N-00W	23	-	0,4	0,15
25.7.1963	14.54	54N-03W	22	-	1,0	0,38
29.7.1963	6.30	53N-12W	20	-	5,0	1,90
3.8.1963	22.42	57N-53W	22	-	1,0	0,38

Anmerkung: Das 1962 eingebaute Gerät hatte eine geringere Auflösung für die vertikale Beschleunigung des Flugzeuges als das 1963 installierte Gerät.
1963 wurde mit größerem Vorschub und größerer Auflösung gearbeitet als 1962. Eichungen für die vertikale Beschleunigungsanzeige fanden nicht statt.

Das durchflogene Luftdruckfeld wurde nicht registriert, so daß auch eine Reihe lokaler Tiefdruckgebilde existieren kann. Die Windregistrierungen weisen auf bestehende Kleinstwirbel. Die Wolkenaufnahmen durch Satelliten der später gestarteten ESSA-Serie zeigen derartige lokale Strömungsgebilde. Die sogenannte " fingerförmige Struktur " des jetstream kann durch eine Reihe dieser kleinen Wirbel oder auch durch Mäander vorgetäuscht werden.

Die raschen Veränderungen in den Höhenwetterkarten, die bei Auftreten starker Windscherungen zu bemerken sind, deuten auf ageostrophische Winde. Inwieweit die Erscheinungen wetterhafter Natur sind, also zeitlich und räumlich eng begrenzt, ist schwer zu entscheiden. Ebenfalls unbeantwortet bleibt die Frage, ob die Bänder (Wirbel) hoher Windgeschwindigkeit in Gebieten mit großräumig geringen Druckunterschieden und schwachen Winden die Allschichten - Zyklonen verlagern oder auffüllen. Sie könnten auch neue Zirkulationssysteme ausbilden und eventuell zu Zyklogenesis führen. Meist tritt eine Aufwölbung von Hochdruckgebilden (wie 26.6.1963) ein. Die Kartenfolge vom 25. bis 27.6.1963 zeigt einen raschen Übergang von einer glatten zonalen Strömung über dem nördlichen Atlantik (25.6.) zu einer Mäanderbildung mit zwei Trögen auf 50° W und 2° W und einer Aufwölbung hohen Druckes am Boden und in der Höhe auf 30° W. Die bei hohen Windunterschieden ablaufenden Umstellungen der Zirkulation sind nicht eindeutig. Die hohen Windunterschiede können auch aus Vorgängen in niederen Schichten folgen. Es sind wahrscheinlich außer den scharfen Randscherungen des jetstream kräftige Initialstörungen erflogen worden. Die Einflüsse niederer Schichten, insbesondere Vertikalbewegungen, auf das obere Windfeld scheinen sehr viel stärker zu sein, als heute allgemein geglaubt wird.

Die geographische Lage der Orte, an denen die hohen vektoriellen Differenzen zwischen zwei benachbarten Windwerten festgestellt wurden, häufen sich in der Nähe der Küsten beider Kontinente. Meßfehler, die durch unterschiedliche Reflektionsstärken über Wasser oder Land entstehen, liegen nicht vor, weil die Veränderungen des Windfeldes nicht über der Küstenlinie erfolgen. Die Lage der ausgedehnten Tröge vor den Kontinenten ist ein meteorologisches Phänomen und ist auch in Mittelkarten zu erkennen (Raethjen (41)).

Die Sicherheitsforderungen im Luftverkehr verlangen eine Diskussion auch derjenigen Fälle, die außerhalb der Grenzen des dreifachen mittleren Fehlers liegen und eine Abschätzung des maximalen Fehlers. Hierbei wendet man sich einer Statistik der Extremfälle zu, die anderen Gesetzmäßigkeiten gehorcht. Die wenigen aufgetretenen Fälle lassen die Abschätzung eines "höchstmöglichen Windunterschiedes auf 100 km" nicht zu.

6. Ein tageszeitlicher Gang im Höhenwindfeld

Zur Feststellung systematischer Fehler der Wetterkarte oder des erflogenen Windes wurden die vektoriellen Differenzen zwischen erflogenen Winden und geostrophischen Winden nach Weltzeit (GMT) geordnet. Überraschenderweise streuen die Werte nicht zufällig um den Nullpunkt. Ein weltzeitlicher Gang im Höhenwindfeld könnte durch die Verlagerung der Druckgebilde bei unglücklicher statistischer Verteilung der Daten vorgetäuscht sein. Die zahlreichen Daten in der Mitte und am Ende des zyklonalen Bogens eines jetstream vermögen eine Rechtdrehung des Windes vorzutäuschen, sofern an den Meßpunkten eine ostwärtige Verlagerung der Tröge stattfindet. Speziell könnte ein vor dem Wetterkartentermin (00 GMT oder 12 GMT) erflogener Südwind in einen Nordwestwind bis zum nächsten Termin überführt werden. Dagegen dürften die nach einem Termin festgestellten Nordwestwinde zur fraglichen Zeit noch Südwestwinde gewesen sein. Falls die Häufung der vorliegenden Daten in Trögen einen wesentlichen Einfluß haben sollten, und diese Tröge rasch nach Osten wanderten, wäre für die Zeiten von 01 GMT bis 06 GMT und von 13 GMT bis 18 GMT zum geostrophischen Wind eine zusätzliche südliche Komponente zu addieren und für die Zeiten von 06 GMT bis 11 GMT und von 18 GMT bis 23 GMT eine nördliche Komponente zu addieren, um den erflogenen Wind zu bekommen. Bild 11 widerlegt diese Vermutung.

Ein Tagesgang im Windfeld ist für mehrere Stationen nachgewiesen worden. Hierbei wurden entweder die Abweichungen vom Tagesmittel für die einzelnen Stunden berechnet oder die tageszeitlichen Schwankungen durch Periodogrammanalyse von vier Radiosonden-Aufstiegen pro Tag ermittelt. Hering und Borden (26) stellen für Shreveport (U.S.A.) für Juli 1958 eine tageszeitliche Schwankung in der Westkomponente des Windes in 12 km Höhe von 4,5 $m\,sec^{-1}$ und in der Südkomponente eine von 5,5 $m\,sec^{-1}$ fest. Das Maximum der südlichen Komponente des Differenzwindes fällt auf den späten Nachmittag, das Maximum der nördlichen Komponente des Differenzwindes (Tagesmittel - Stundenmittel) liegt zwischen 3 und 6 Uhr Ortszeit. Der höchste zusätzliche Westwind herrscht um Mitternacht; der zusätzliche Ostwind erreicht sein Maximum gegen Mittag.

Für die Azoren stellen Harris, Finger, Teweles (25) Höchstwerte der Windvariation in der 200-mb-Fläche fest. Die Amplituden betragen im Jahresmittel für die östliche Windkomponente in der Tagesschwingung 0,56 m sec^{-1}, in der Halbtagesschwingung 0,26 m sec^{-1} und für die nördliche Windkomponente in der Tagesschwingung 0,52 m sec^{-1} und in der Halbtagesschwingung 0,08 m sec^{-1}. Eine bedeutende Tagesschwingung weisen beide Komponenten auf, eine bedeutende Halbtagesschwingung ist nur für die östliche Komponente festzustellen.

Die zitierten Ergebnisse über tageszeitliche Schwankungen des Windfeldes beschreiben die Abweichungen vom Mittelwert. Dagegen zeigen die tageszeitlichen Unterschiede zwischen erflogenen und geostrophischen Winden Kräfte an, die zusätzlich zum analysierten Druckfeld wirksam sind. Diese Kräfte dürften das Wetter beeinflussen. Zwar sind die den erflogenen Winden zugeordneten geostrophischen Daten zeitlich nicht korrigiert worden, jedoch erzielen die Ergebnisse erste Anhaltspunkte zur Adataption zwischen Druck- und Windfeld, insbesondere für die Termine 00 GMT und 12 GMT (Bild 11). Ein geringer Anteil der nachgewiesenen "ageostrophischen Komponenten" dürfte durch großräumige Veränderungen des Druckfeldes in der Zeit zwischen Messung mit dem Flugzeug und Radiosondenaufstieg verursacht sein. Bessere Aufschlüsse zur Adataption gäben gleichzeitige Registrierungen des Druck-, Temperatur- und Windfeldes in mehreren Flugzeugen.

In Bild 12 sind die Vektoren gezeichnet, die zur jeweiligen Ortszeit dem geostrophischen Wind zu addieren sind, um den erflogenen Wind zu erhalten. Die Spitzen dieser Vektoren sind mit . gezeichnet und mit der Uhrzeit (Ortszeit) versehen.

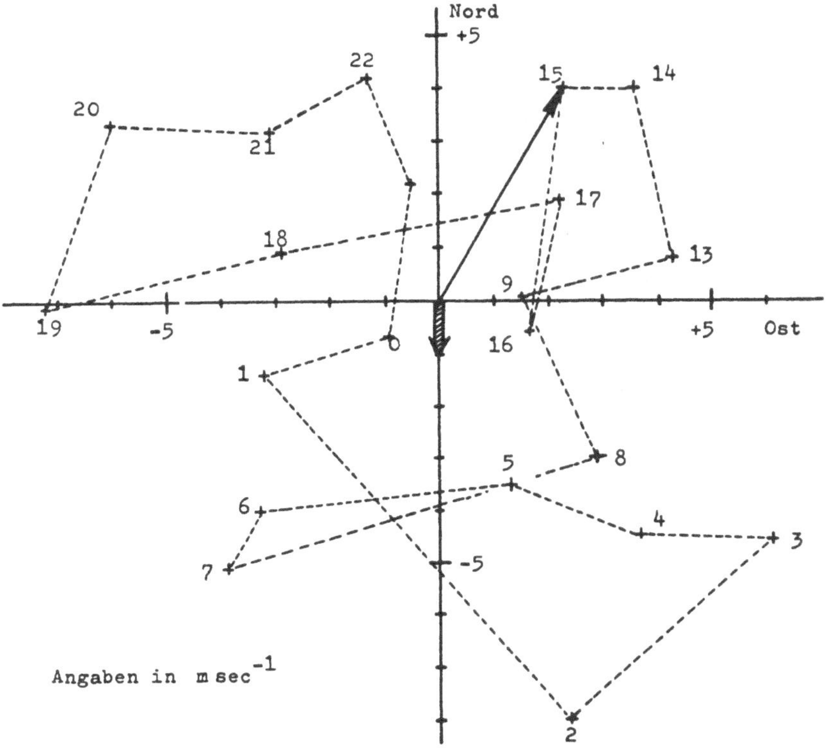

Bild 11: Weltzeitlicher Gang der ageostrophischen Komponente

Anmerkung: Die Spitze des geostrophischen Windvektors liegt im Nullpunkt (Mittelpunkt) des Koordinatensystems. Die mit Ziffern bezeichneten Marken (+) stellen die Spitze eines ageostrophischen Vektors dar, der dem geostrophischen Windvektor zu addieren ist, um den erflogenen Windvektor zu erhalten. Dieser ageostrophische Windvektor führt vom Ursprung des Koordinatensystems zur Marke, deren Ziffer die Uhrzeit (hier in GMT , Weltzeit) angibt. Der fette Pfeil gibt den mittleren ageostrophischen Wind wieder. Der dünne Pfeil kennzeichnet den mittleren ageostrophischen Windvektor der um 15 GMT dem mittleren geostrophischen Wind zu addieren ist, um den mittleren erflogenen Wind zu erhalten. Es ist ein zusätzlicher, ageostrophischer Südwestwind.

Sie gehen vom Nullpunkt (Schnittpunkt der Achsen im Koordinatensystem) aus. Ein Vektor steht jeweils für das Mittel aller Werte, die zwischen der bezeichneten Stunde und der nächstfolgenden gemessen wurden. Kollektive mit weniger als 20 Wertepaaren blieben unberücksichtigt. Bild 13 zeigt die Verteilung der Vektorspitzen des ageostrophischen Windes, wenn alle Unterschiede zwischen erflogenen und geostrophischen Winden, die größer als 20 m sec^{-1} (vektoriell) sind, ausgeschlossen werden. Die Lage der ageostrophischen Vektoren verändert sich nicht grundlegend.

Die Aufspaltung nach Ortszeit ergibt einen tageszeitlichen Gang der ageostrophischen Winde und zwar ist am Tage zwischen 9 Uhr und 19 Uhr zu dem geostrophischen Wind (aus der Wetterkarte entnommen) eine zusätzliche südliche Komponente und in der Nacht zwischen 20 Uhr bis 7 Uhr eine zusätzliche nördliche Komponente zu addieren, um den erflogenen Wind (Mittel) zu erhalten. Dem Tagesmittelwert des geostrophischen Windes ist ein Nordwind von 1 m sec^{-1} hinzuzufügen, um das Tagesmittel der erflogenen Winde zu bekommen. Die zonalen Komponenten (Ost - West) addieren sich zu Null.

In das Material wurden keine systematischen Fehler durch das Fluginstrumentarium hineingetragen. Die mittleren Unterschiede zwischen erflogenen und geostrophischen Winden haben ihre Ursache im tageszeitlichen Gang des Temperatur- und Druckfeldes, sowie im Austausch der Luftmassen zwischen den Tropen und der Polarkalotte. Diese ageostrophischen Winde sind als Beitrag in dem "mittleren Fehler des geostrophischen Windes nach der Wetterkarte" enthalten, der im Abschnitt 4 abgeleitet wurde.

Die harmonische Analyse erbringt bei Aufteilung der Daten nach Ortszeit für die Nordkomponente der vektoriellen Differenz zwischen erflogenen und geostrophischen Winden nur eine ganztägige Schwingung. Die Amplitude der Schwingung beträgt 3,9 m sec^{-1}. Das Maximum tritt um 3 Uhr auf. Die von anderen

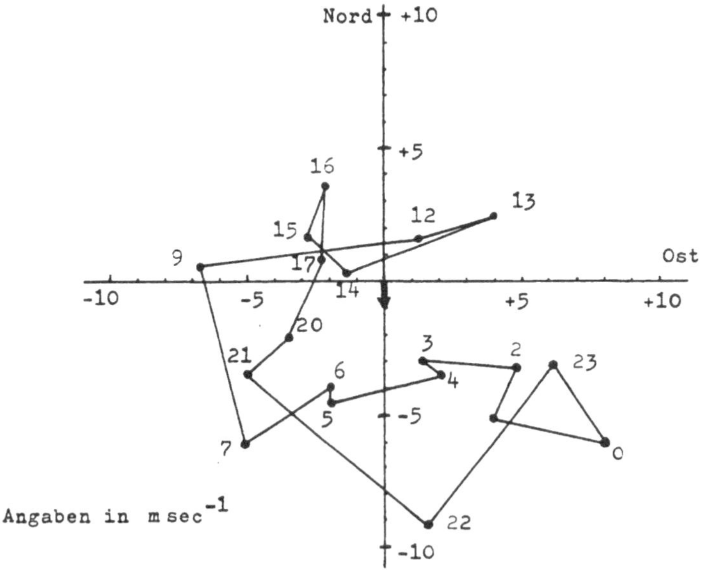

Bild 12: Ortszeitlicher Gang des ageostrophischen Windes

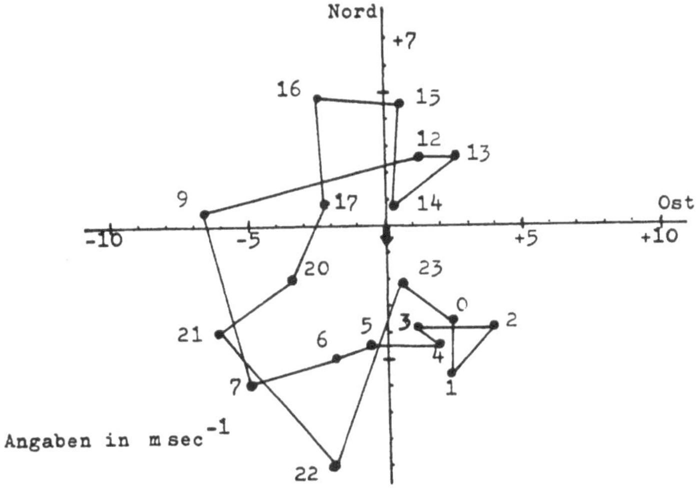

Bild 13: Ortszeitlicher Gang des ageostrophischen Windes
unter Ausschaltung grober Fehler in der Wetterkarte

Autoren festgestellten untergeordneten Perioden im Windfeld, wie zwölfstündige Schwingung, übersteigen bei den ageostrophischen Winden in der Nordkomponente nicht den allgemeinen Störpegel von $1\,m\,sec^{-1}$. Für die Ostkomponente lassen sich Tages-, Halbtages- und Vierteltagesschwingungen nachweisen, deren Amplituden ebenfalls überzufällig sind. Sie betragen $2{,}8\,m\,sec^{-1}$ für die 24-stündige, $3{,}6\,m\,sec^{-1}$ für die 12-stündige und $2{,}4\,m\,sec^{-1}$ für die 6-stündige Schwingung.

Die ganztägige Schwankung in der Nordkomponente des ageostrophischen Windes dürfte durch den tageszeitlichen Temperaturgang, verbunden mit einem tageszeitlich verschiedenen Austausch zwischen Tropen und Polargebieten, verursacht sein. Diese Austauschvorgänge sind aber nicht durch einfache Schwankungen in den Nord- oder Südströmungen zu beschreiben, die in der Stratosphäre der allgemeinen westlichen Strömungen überlagert sind. Vielmehr führen die für die Tagesstunden festgestellten südlichen Komponenten des ageostrophischen Windes zu einer Vermehrung der Impulszufuhr in der Westströmung und die für die Nachtstunden festgestellten nördlichen Komponenten des ageostrophischen Windes zu einer Verminderung der Impulszufuhr in der Westströmung der freien Atmosphäre.

Raethjen (41) erklärt die Entstehung starker Inversionen durch Abgleitvorgänge in den darüber liegenden Schichten. Die Tropopause ist die stärkste Temperatur - Inversion der Atmosphäre und verdankt ihre Entstehung ähnlichen Vorgängen. Das ausgewertete Material kann die Existenz abgleitender Luftmassen nicht beweisen. Es bietet Einblicke in den Wirkungsmechanismus vorhandener Kräfte (ageostrophischer Winde), die im Tropopausenniveau notwendige Strömungen polwärts ausbilden können.

Die statistische Sicherheit im vorhandenen Datenkollektiv reicht nicht aus, verbindlich zu entscheiden, ob in der Stratosphäre eine stärkere polwärts gerichtete ageostrophische Strömung als in der Troposphäre existiert (siehe mittlere Nordkomponente des erflogenen Windes, Bild 5, Seite 70). Anzeichen einer stärkeren ageostrophischen polwärtsgerichteten Strömung sind besonders am Tage für die Stratosphäre mehr als für die Troposphäre vorhanden. Derartige Strömungen führen durch die Schrumpfung der Atmosphäre polwärts zu Abgleitvorgängen. Direkt oberhalb der Tropopause ist in den Mittelwerten die höchste südliche und auch westliche Komponente des erflogenen Windes festzustellen.

Gezeiten, die durch Hebungen beziehungsweise Senkungen der Druckflächen mit der tageszeitlichen Aufheizung entstehen, bewirken (zusätzlich zu den tageszeitbedingten Schwankungen in der Impulszufuhr des Westwindes) das komplizierte Schwingungsbild in den zonalen Komponenten, worauf nicht näher eingegangen werden soll. Es war lediglich Ziel, auf periodische Abweichungen zwischen Druck- und Windfeld über dem Nordatlantik hinzuweisen.

Der Tagesgang des ageostrophischen Windes wird nicht für alle Jahreszeiten gleich sein. Insbesondere werden die Amplituden im Winterhalbjahr geringer sein. Der ageostrophische Wind mit nördlicher Komponente während der Nachtstunden und mit südlicher Komponente während des Tages liegt für das untersuchte Material außerhalb der Zufallsgrenzen.

Das Luftdruckfeld ist dem erflogenen Windfeld nicht zeitlich angepaßt worden, so daß die beschriebenen "ageostrophischen Winde" sich aus den tageszeitlichen Schwankungen des erflogenen Windfeldes gegenüber dem geostrophischen Windfeld von 00 GMT beziehungsweise 12 GMT zusammensetzen.

7. Bemerkungen zu ausgewählten Registrierergebnissen

7.1 Vergleich der Registrierergebnisse vom 6.4.1962 mit der Wetterlage

7.1.1 Wetterlage vom 6.4.1962

Die nordhemisphärische Zirkulation zeigt in der 200-mb-Fläche an jenem Tage vier Tröge mit abgeschlossenen Höhen-Tiefdruckzentren und zwar über West-Europa (10° E), Sibirien (120° E), Alaska (150° W) und Labrador (60° W). Über dem östlichen Atlantik wölbt sich ein Hochdruckkeil. Aus dem europäischen Trog schnürte sich ein Kaltlufttropfen ab, der in den nächsten Tagen bis zum Schwarzen Meer vordrang.

Der polare Ast des Strahlstromes bildet über Labrador bis Süd-Grönland ein Starkwindband aus. Der subtropische Ast ist über dem mittleren Atlantik stärker ausgebildet und reicht über Europa bis in das Mittelmeer.

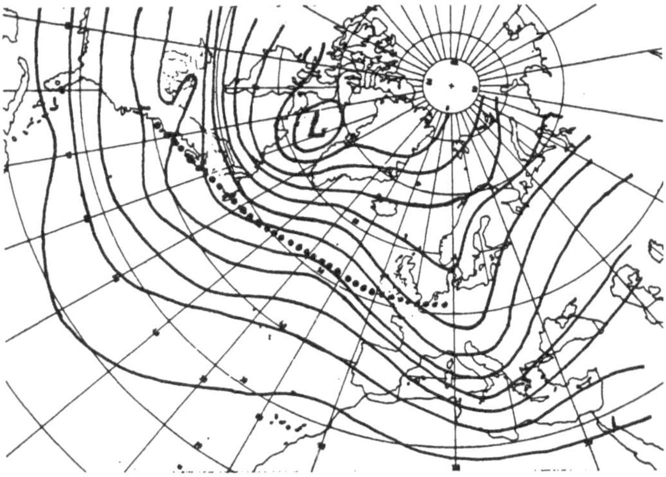

Bild 14: Wetterlage 6.4.1962 - 00 GMT in 200 mb
Linien: Punktiert = Flugverlauf, ausgezogen = Isohypsen

7.1.2 Streckenführung des Fluges

Der Flug LH 403 am 6.4.1962 führte von New York nach Köln. Er folgte über dem atlantischen Ozean dem antizyklonalen Strömungsbogen nördlich des subtropischen Azoren-Hochdruckgebietes. Der Flug blieb hart südlich eines Tiefdruckgebietes, dessen Kern am Boden südöstlich Kap Farvel lag und dessen Frontensysteme überflogen wurden. Der Flug endete in dem stark ausgeprägten Trog mit einem Kaltlufttropfen über Mitteleuropa.

Die Tropopause wurde zweimal isobar geschnitten. Der erste Teil des Fluges erfolgte bis 57° W in der Troposphäre, nach einem Steigflug um 2000 ft bis ungefähr 35° W in der Stratosphäre, der anschließende Teil bis über Irland in der oberen Troposphäre und der letzte Abschnitt in der Stratosphäre. Der erste isobare Schnitt der Tropopause lag auf der subtropischen Seite des Strahlstromes, der zweite dagegen in Stromrichtung links vom Kern des Starkwindfeldes. Die Tropopause wurde bei 35° W über einer Boden-Kaltfront geschnitten (Lage der Kaltfront nach Analyse der Bodenkarte).

7.1.3 Analyse der Registrierergebnisse und Radiosondenwerte

Bild 15 (Vertikalschnitt längs des Flugweges, Seite 110 stellt die während des Fluges registrierten Daten mit den Messungen der Radiosonden zusammen. Die fettgezogene Linie kennzeichnet den Verlauf der Tropopause längs des Fluges. Über dem westlichen Atlantik liegt nach Flugzeugmessungen die Tropopause (gestrichelt) über dem Flugniveau (amerikanische Radiosonden abweichend). Die Flughöhe ist als dünne Gerade eingezeichnet. Die nahezu horizontal verlaufenden dicken Linien geben die Isentropen-Schnittflächen wieder. In Richtung der Ordinate sind als dünne Striche die Radiosondenaufstiege markiert. Im oberen Teil des Bildes sind die Registrierergebnisse zusammengestellt. Der Kurvenverlauf bezeichnet die registrierte Temperatur. Die Strecken mit registrierter Flugunruhe sind als fette Balken bezeichnet. Die kurzen Windpfeile geben den erflogenen Wind für einen zeit-

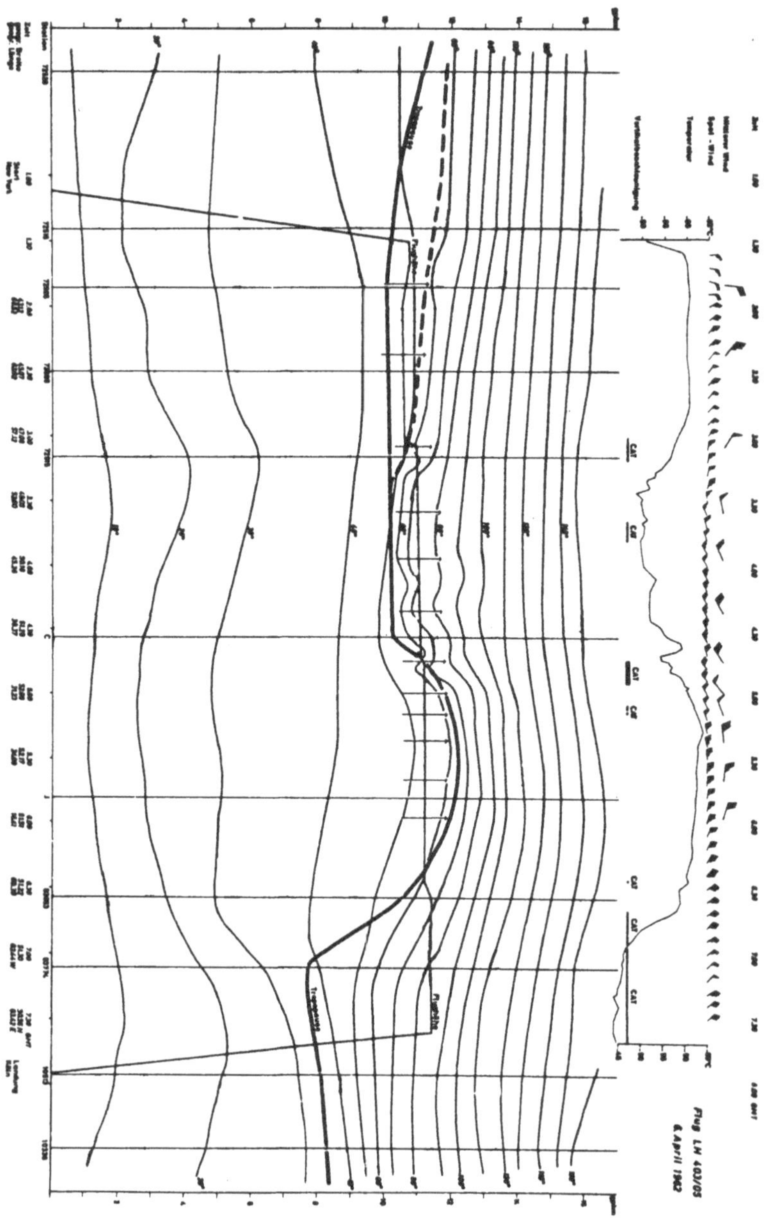

Bild 15: Vertikalschnitt längs des Flugweges am 6.4.1962

lichen Abstand von 6 Minuten an. Die großen Windpfeile bedeuten mittlere Winde, die aus den über längere Intervalle integrierten Kurven des Schreibstreifens errechnet wurden. Die mit dem Flugzeug gemessenen Temperaturen und Winde stimmen gut mit den Ergebnissen der Radiosonden-Aufstiege überein, wie Bild 15 zeigt.

Die Differenzvektoren zwischen den Wind-Mittelwerten, die durch Planimetrierung der registrierten Kurven über eine Streifenlänge entsprechend ungefähr 20 Minuten Flugzeit (je nach Ablesung der Höhenmesser im Flugzeug) gebildet wurden, und den im Abstand von 6 Minuten direkt gemessenen (nicht ausgeglichen) Winden sind in Bild 16 zusammengestellt. Die Streckenlänge, über welche gemittelt wurde, ist durch die während des Fluges erfolgten Ablesungen der Instrumente vorgegeben. In Bild 16 sind die Differenzvektoren in einem geographisch-orientierten Koordinatensystem dargestellt (Nord nach oben, Ost nach rechts). In der Figur stellt der Abstand zwischen Nullpunkt und einer Marke (+,x) den Betrag des Differenzvektors dar. Die Veränderungen des Windes, hier als Differenz zwischen Zwanzig-Minuten-Mittel und Einzelwert gegeben, sind überwiegend kleiner als 10 m sec^{-1}; sie erreichen vereinzelt mehr als 20 m sec^{-1} (äußerer Kreis in Bild 16). Für die Wetterkarte sind die mittleren Winde besser geeignet als die Einzelwerte. Die den Flugwetterwarten übermittelten erflogenen Winde (spot-winds) werden über das KIFIS und Doppler-Radar errechnet und sind dadurch ebenfalls über eine kurze Strecke geglättet.

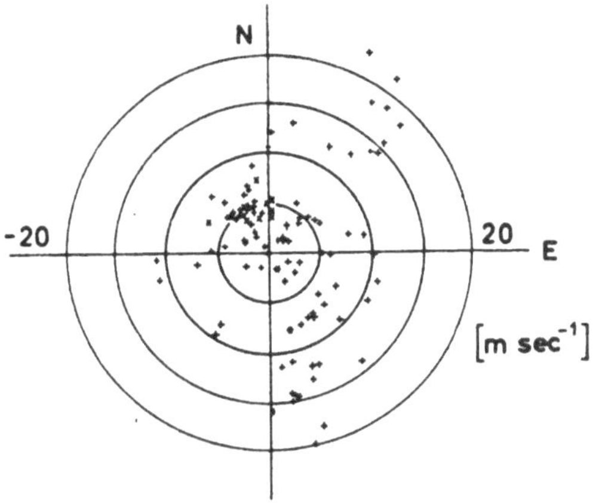

Bild 16:
Punktverteilung der Differenzvektoren zwischen erflogenen Winden und ihren Mittelwerten

Beispiel vom 6.4.1962

Marken:
+ Stratosphäre
x Troposphäre

In Bild 17 sind die Querwind-Komponenten (oben) sowie die registrierten Temperaturen (unten) des Fluges vom 6.4.1962 dargestellt. In der Abszisse steht die geographische Länge, in der Ordinate ist die Querwind-Komponente, d.h. die Komponente des Windes senkrecht zum Kurs des Flugzeuges über Grund aufgetragen. Für die Darstellung der Querwind-Komponenten gilt die Skala auf der linken Seite der Zeichnung. Die Temperatur-Skala steht auf der rechten Seite der Zeichnung.

Dünn ausgezogen ist die Querwind-Komponente, die sich aus den Doppler-Daten des Fluges ergibt (Q_f).
Gestrichelt ist die Querwind-Komponente des geostrophischen Windes, der sich aus den Meßdaten des Flugzeuges errechnet (Q_r).
Punktiert ist die geostrophische Querwind-Komponente aus Daten der Wetterkarte (Q_k).
Fett ausgezogen ist die Differenz zwischen Doppler-Querwind Q_f und geostrophischem Querwind Q_r (beides Flugzeugdaten).
Fett punktiert ist im unteren Drittel des Bildes die registrierte Temperatur dargestellt.

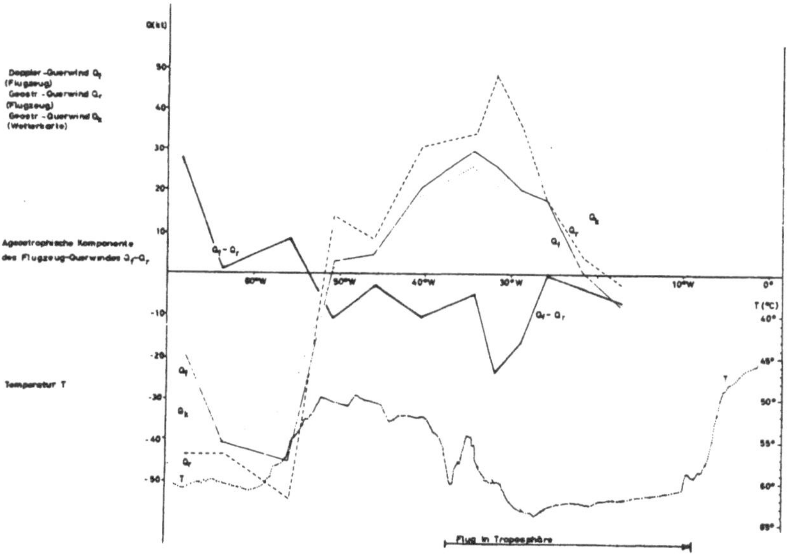

Bild 17: Querwind-Komponenten und Temperatur am 6.4.1962

Es herrschten am 6.4.1962 auf der polaren Seite des Subtropikjetstream in zyklonaler Krümmung während des Fluges untergeostrophische Winde, dagegen auf der antizyklonalen Scherungsseite (gleichzeitig in antizyklonaler Krümmung, südlich des Strahlstrom-Kernes) übergeostrophische Winde. Der jetstream wird in jenen Tagen nach Süden zur antizyklonalen Scherungsseite gedrängt.

Die ageostrophische Querwindkomponente (Differenz Q_f - Q_r) ist während des Tropopausen-Schnittes auf 35° W (im Bild 17 an der Stelle mit niedriger Temperatur, vor kurzzeitigem neuem Anstieg) im Windmaximum niedrig. Die größten ageostrophischen Komponenten des Querwindes treten im später durchflogenen kälteren Teil der Troposphäre auf. Die kleinen Unterschiede in der fett ausgezogenen Kurve bei 40° W sind nicht gesichert.

Es kann nicht auf Grund dieser Registrierung und protokolliertem Verlauf der geometrischen Höhe des Flugzeuges gefolgert werden, daß beim Durchflug der Tropopause das kältere Gebiet mit höheren Windgeschwindigkeiten auch größere ageostrophische Winde als das wärmere Gebiet zeigt. Eine kontinuierliche Registrierung der Querwindkomponenten würde bessere Aufschlüsse liefern. Die maximale Geschwindigkeit des geostrophischen Querwindes wird zeitlich später als die des Doppler-Querwindes zu Beginn des troposphärischen Flugteils erreicht (beide liegen durch Anlage des Fluges zeitlich später als das Windmaximum). Nach Bild 14 (Wetterlage vom 6.4.1962) liegt das Windmaximum auf der polaren, der stärkste Druckgradient aber auf der erst später angeflogenen subtropischen Seite des Strahlstromes.

Bild 18 zeigt den Verlauf von Temperatur und Wind bei isobarem Schnitt der Tropopause nach Registrierergebnissen vom 6.4.1962. Auf der Abszisse ist die Flugzeit in GMT angegeben. Die Ordinate gibt den Betrag der Windgeschwindigkeit in Knoten sowie die Temperatur in °C wieder (tiefe Temperaturen oben!). Im unteren Teil des Bildes wird die Windrichtung dargestellt.

Der Flug erfolgte (im Bild von links nach rechts) anfangs in der Stratosphäre und ab 4.40 GMT in der Troposphäre. Beim Einflug von der Stratosphäre in die Troposphäre wurde ein markanter Abfall, Wiederanstieg und weiterer Abfall der Temperatur sowie ein scharfer Windsprung registriert. Der Wind fluktuierte stärker im letzten Teil des stratosphärischen Fluges. Die Windgeschwindigkeit nimmt vor dem Durchflug der Tropopause stark ab und steigt dann rasch während des Schnittes auf Höchstwerte von 42 m sec^{-1}, die mit den tiefsten registrierten Temperaturen zusammenfallen. In der hohen Troposphäre wird anschließend etwas wärmere Luft durchflogen (hier könnte noch einmal stratosphärische Luft angeschnitten sein). Die Windgeschwindigkeit erniedrigt sich dort rasch. In der oberen Troposphäre bleiben die Fluktuationen des Windes stark, gleichzeitig ist der Flug unruhig (CAT). Nach 300 km Flugweg in der Troposphäre nehmen die Fluktuationen des Windes und die Flugunruhe (CAT) ab. Der Flug verlief bis zum Einflug in den Kaltlufttropfen über Europa ruhig.

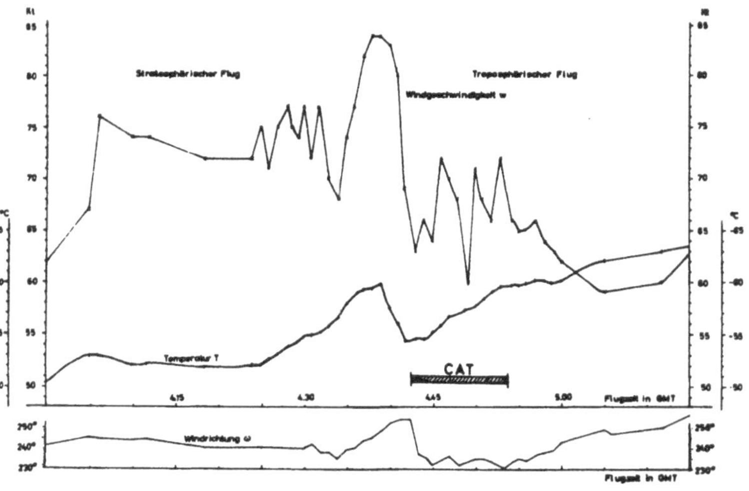

Bild 18: Verlauf von Temperatur und Wind bei isobarem Schnitt der Tropopause am 6.4.1962

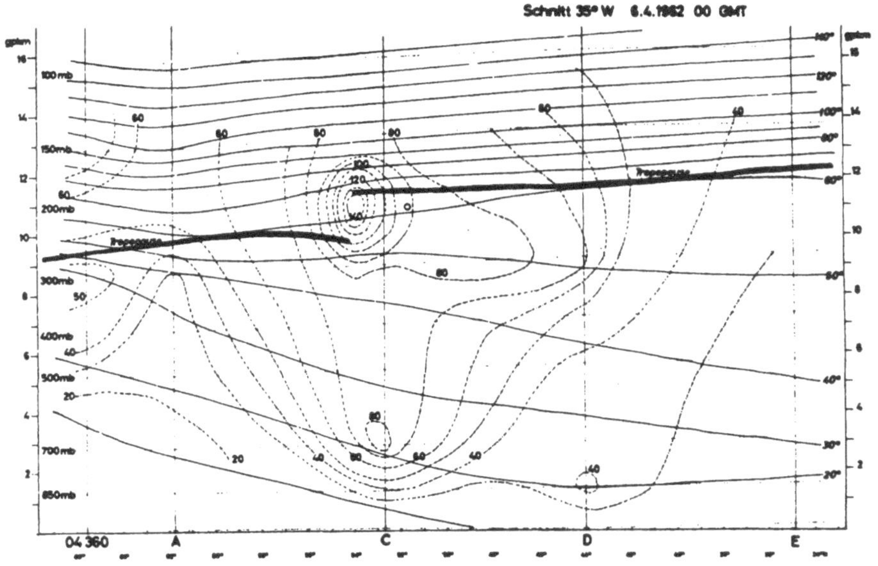

Bild 19: Vertikaler Schnitt auf 35° West am 6.4.1962 - 00 GMT

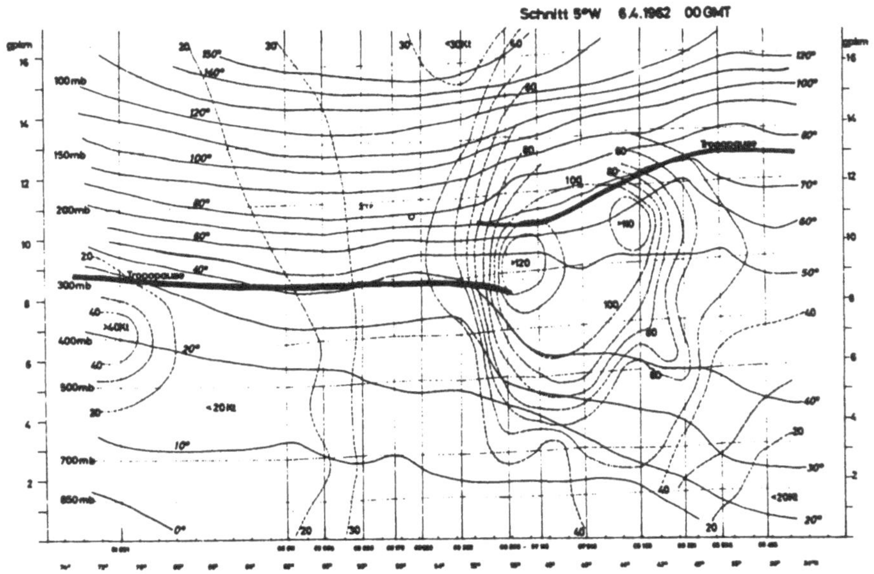

Bild 20: Vertikaler Schnitt auf 5° West am 6.4.1962 - 00 GMT

Die Temperaturschwankung betrug an der Tropopause auf 35° W im isobaren Schnitt ungefähr 10° . Bei dem Wiedereinflug der Maschine in die Stratosphäre auf der östlichen Seite des atlantischen Hochdruckrückens wurden keine großen Temperaturschwankungen registriert. Die Temperatur stieg dort stetig von -59° auf -46° in 24 Minuten Flugzeit. Die Tropopause fiel über England weit unter das Flugniveau ab, so daß hohe Stratosphären-Temperaturen registriert wurden. Über dem westlichen Atlantik flog die Maschine stets nahe der Tropopause.

Die hohen Temperaturschwankungen direkt an der Tropopause auf der westlichen Seite des antizyklonalen Strömungsbogens mag durch nächtliche Ausstrahlung und durch abgleitende Bewegungen stratosphärischer Luftmassen oberhalb der Tropopause begünstigt worden sein. Vom südlichen Nordatlantik in Höhe des Wetterschiffes E ist über die Wetterschiffe D und C bis nach Grönland in der Stratosphäre ein Gefälle der Isentropen-Flächen zu verfolgen, wie Bild 19 in einem Vertikalschnitt von Nord (links) nach Süd auf 35° W andeutet. Auf der östlichen Seite des subtropischen Hochdruckrückens ist in der Stratosphäre am Standort des Flugzeuges kein Gefälle der Isentropen-Flächen mit der Luftströmung festzustellen. Der Abfall erfolgt südlicher bei einer Strömung mit nördlicher Komponente (Aufgleiten). Bild 20 zeigt einen Vertikalschnitt von Nord nach Süd auf 5° W . Der jetstream liegt nach dieser Analyse südlich des Flugzeug-Standortes auf 5° W . Der Standort des Flugzeuges lag auf 35° W nördlich des Strahlstromkernes.

Die Temperatur-Registrierung zeigt beim isobaren Schnitt der Tropopause auf 35° W das Nebeneinanderliegen einer örtlichen Isentropensenkung (Isentropengraben) auf der troposphärischen Seite und einer örtlichen Isentropenhebung (Isentropenbuckel) auf der stratosphärischen Seite des Fluges.

Der Wind dreht mit einsetzenden tiefen Temperaturen recht und springt an der tiefsten Stelle der nachfolgenden Isentropensenkung ungefähr in die frühere Richtung zurück.

7.2 Vergleich der Registrierergebnisse mit der Wetterlage vom 26. bis 28.3.1963

7.2.1 Wetterlage vom 26.3.1963 bis 28.3.1963

Am 26.3.1963 bis 28.3.1963 herrscht eine starke Westdrift, die vom östlichen Pazifik über Nord-Amerika und den Atlantik bis nach Rußland reicht. Bei dieser spätwinterlichen Wetterlage gruppieren sich auf der Nordhalbkugel in der 200-mb-Fläche um den Pol mehrere Tröge, die über Baffin-Land, Alaska, Ost-Sibirien und dem Ural liegen.

Der jetstream weist in der 200-mb-Karte Winde über 60 m sec^{-1} auf. Die Höhenströmung in 200 mb ist in ein nördliches und ein südliches Starkwindband über den Vereinigten Staaten und dem Nordatlantik geteilt (Bild 21). Diese Teilung wird durch Tiefdrucksysteme über dem Huron-See und dem Staate Washington begünstigt.

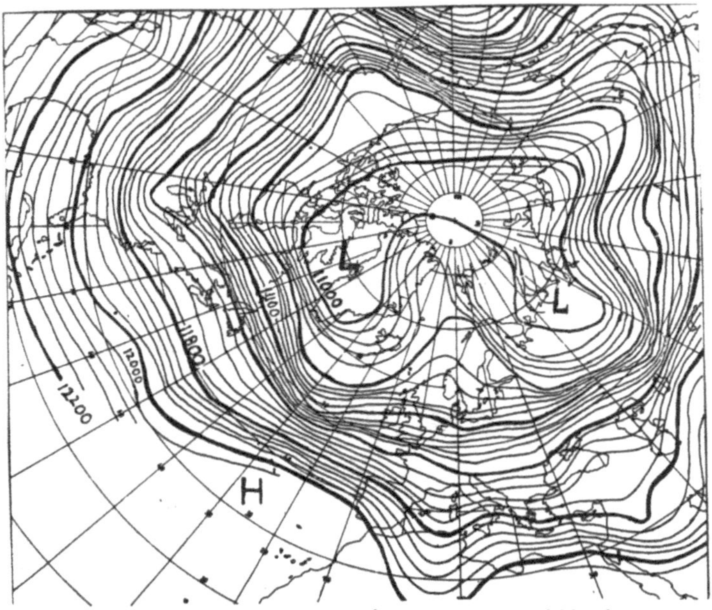

Bild 21: Wetterlage vom 27.3.1963 - 00 GMT , 200 mb

Die sehr kräftige Frontalzone steuert eine Reihe von Zyklonen nach Osten. Die erste einer Reihe ist über die Ostsee nach Rußland gezogen, auf ihrer Rückseite strömt Kaltluft nach Norddeutschland. Das nordatlantische Zentraltief bewegt sich langsam von der Dänemarkstraße in östlicher Richtung. In einem breiten Strom dringt bei Grönland Polarluft ostwärts auf den Atlantik vor. Ein Randtief liegt am 27.3.1963 um 00 GMT über der Mündung des St.Lorenz-Stromes (Überflug), ein kräftigeres folgt über den Großen Seen nach. Weitere Tiefdruckgebiete bestimmen das Wetter der Westküste Amerikas. Sie liegen auf dem nordöstlichen Pazifik mit den Fronten ungefähr parallel zur Küste. Ein Hochdruckgebiet von 1030 mb befindet sich auf 40° Breite. Ein subtropisches Hoch schließt sich westlich des nordamerikanischen Kontinentes an.

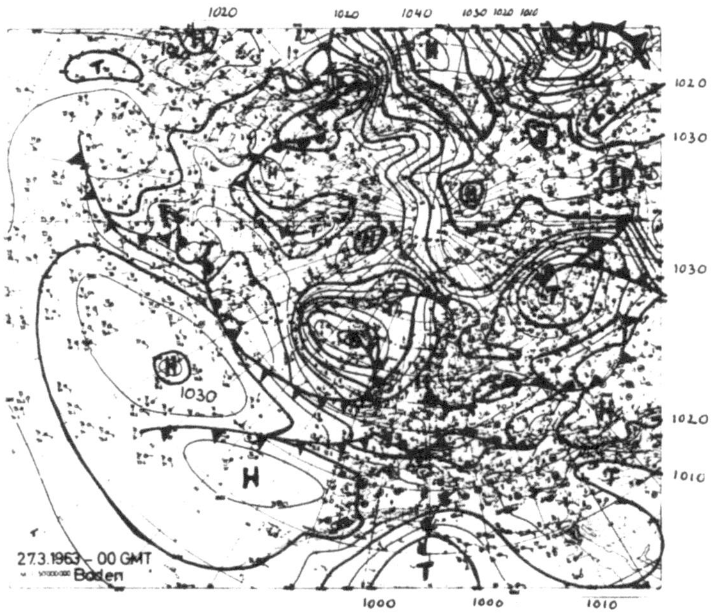

Bild 22: Wetterlage vom 27.3.1963 - 00 GMT , Boden
(Aus:Deutscher Wetterdienst, Tägl.Wetterbericht)

7.2.2 Streckenführung des Fluges

Der Flug führte von Frankfurt über London, Nord-Irland bis zum 59. Breitengrad südlich Grönlands und weiter über Wetterschiff B , Labrador nach Montreal. Auf der Teilstrecke bis Grönland wurden die hohen Gegenwindgeschwindigkeiten auf der Nordseite des Strahlstromes umflogen. Der Flug kreuzte erst anschließend über Kanada den polaren Zweig des Strahlstromes. Der Flug verlief über dem Atlantik in der Stratosphäre und über dem amerikanischen Kontinent in der Nähe der Tropopause.

Auf der weiteren Strecke von Montreal über Denver nach San Francisco wurde der subtropische Zweig des jetstream durchflogen. Bis zum 100. Längengrad befand sich die Maschine in der Stratosphäre. Die Tropopause ist in der Nähe des Strahlstrom-Kernes geschnitten worden. Der anschließende Teil des Fluges lag in der Troposphäre. Die Flughöhe betrug über den U.S.A. 39 000 ft .

Der Rückflug erfolgte in 41 000 ft anfangs in der Troposphäre. Das Flugzeug durchflog bei 97° West die Tropopause. Der weitere Flug, nördlich des subtropischen jetstream, blieb in der Stratosphäre.

Der Flug hielt sich zwischen Montreal und London nördlich der stärksten Westwinde. Seine nördlichste Position erreichte er auf 40° West mit 54° Nord. Die Troposphäre wurde zwischen Neu-Fundland und 38° West angeschnitten. Die Flughöhe betrug 36 000 ft. Die atmosphärischen Erscheinungen waren an den niedrigen Tropopausen schwach.

7.2.3 Analyse der Registrierergebnisse und Beobachtungen

Im Gebiet südlich Grönlands lag am 26.3.1963 die Tropopause sehr tief:

Wetterschiff	A	62,0 N	33,0 W	425 mb
Wetterschiff	B	56,5 N	51,0 W	269 mb
Wetterschiff	C	57,0 N	35,5 W	385 mb
Wetterschiff	I	58,8 N	18,8 W	426 mb

Die Bilder 23 bis 26 zeigen Wolkenaufnahmen südlich Islands und Grönlands. Auf Bild 24 ist die Südspitze Grönlands und die Grenze des Eises zwischen Grönland und Island zu erkennen. Die Bilder 23 und 26 sind nach Süden, die Bilder 24 und 25 nach Norden photographiert.

Die Quellbewölkung an der Kaltfront westlich Irlands ging in eine geschlossene Stratocumulus- und Altocumulus-Decke über, die stellenweise kleinere wirbelartige Strukturen erkennen ließ (Bild 23). Nach Norden lockerte sich die Decke in der Nähe des Tiefs bei Island auf. Über den kalten Wassermassen des Ostgrönland-Stromes (Bild 24) und des Labrador-Stromes (Bild 25, links oben) lösen sich die Wolken auf der Nordseite des hochreichenden isländischen Zentraltiefs vollständig auf. In der Nähe des Wetterschiffes B ordnete sich die Stratocumulus-Bewölkung straßenförmig an. Wie in Bild 25 und 24 zu erkennen, laufen die Ketten der zum Teil stärker quellenden Wolken aus der Davis-Straße mit den Isohypsen des Island-Tiefs gebogen bis in den Bereich des Golfstromes, wo sie dann in eine geschlossene Decke übergehen. Der Einfluß des Untergrundes ist für die Ausbildung dieser Wolkenstraßen nicht entscheidend, viel mehr Bedeutung hat das vertikale Windprofil. Der Radiosondenaufstieg des Wetterschiffes B zeigt ein niedriges Windmaximum ($35\,m\,sec^{-1}$) in 700 mb. In den Bildern 24 und 25 ist auch zu erkennen, daß der Abstand zwischen den Straßen um so größer ist, je höher die Cumuli quellen.

Bild 23: Wolken 30° W 26. 3. 1963

Bild 24: Wolken südlich Grönlands (nach NE) 26. 3. 1963

Die Stratosphäre erscheint tief dunkelblau und ist auch gegen Süden ohne Konturen. Die braunrote Tropopause begrenzt die weißblaue Troposphäre am südlichen Horizont gegen das tiefe Blau. Während des Fluges in der Stratosphäre über dem östlichen Atlantik sind am 26.3.1963 (nachmittags) keine Cirren zu beobachten (Bilder 23 - 25). Erst mit Ansteigen der Tropopause vor Labrador (Bild 26) werden schleierartige Konturen in tieferen Niveaus und Cirren am Horizont erkennbar. Das Aufkommen eines dichten Cirrusstratus behindert vor Labrador die Bodensicht. Dieses Wolkensystem gehört zur Aufgleitbewölkung der Vorderseite eines nordamerikanischen Tiefdrucksystems. Über dem amerikanischen Kontinent nimmt nach Eintauchen des Flugzeuges in die Troposphäre die Sicht merklich ab. Die Windgeschwindigkeiten lagen im Flugniveau bei 20 m sec^{-1} aus westlichen Richtungen.

Auf der Strecke Montreal - San Francisco wird zunächst in der Stratosphäre nördlich des subtropischen Jet-Kerns geflogen. Die Tropopause wird auf 41° N und 100° W in Nähe des Jet-Kernes geschnitten. Der anschließende Flug verläuft in der Troposphäre. Mit Annäherung an die Tropopause sinkt die Temperatur von -54,0°C bis -59,7°C (Isentropenbuckel auf stratosphärischer Seite), steigt innerhalb von zwei Minuten auf -54,5°C (Isentropengraben auf troposphärischer Seite) und fällt erneut während des Fluges in der oberen Troposphäre auf der subtropischen Seite des jetstream auf -61,5°C innerhalb der folgenden zwanzig Minuten.

Flugunruhe (CAT) tritt mit Einflug in den Kältemantel der Troposphäre plötzlich und heftig auf. Sie zwingt zur Verminderung der Fahrt. Nach kurzer Beruhigung folgt während des Fluges durch den "Isentropengraben" (wärmeres Gebiet) eine neue Periode mit Clear Air Turbulence (CAT). Der anschließende troposphärische Flug verläuft verhältnismäßig ruhig. Die Windgeschwindigkeit hat ihr Maximum während des Tropopausenschnittes im "Isentropenbuckel" (auf stratosphärischer Seite, verbunden mit den niedrigsten Temperaturen). In der Troposphäre ist in Flugrichtung ein "Rückdrehen" (Drehen nach links) des Windes zu beobachten.

Bild 25: Wolken südlich Grönlands (nach NW) 26.3.1963

Bild 26: Wolken östlich Labrador (nach S) 26.3.1963

Die im stratosphärischen Teil des Fluges am 26./27.3.1963 von Montreal nach San Francisco aufgetretene Flugunruhe (CAT) war eng an eine Isentropenfläche gebunden. Die vertikale Ausdehnung der Turbulenzgebiete war so gering, daß durch kleine Höhenänderungen das Flugzeug wieder in eine ruhigere Fläche gesteuert werden konnte. Im Bereich der Rocky Mountains traten durch Leewellen heftigere Turbulenzen in allen Flugniveaus auf.

Auf dem Rückflug San Francisco - Montreal - London - Frankfurt wurden über den Rocky Mountains Gebiete mit sehr starker Turbulenz in großer Höhe überflogen (Flughöhe teilweise 41 000 ft, 180 mb). Zwischen San Francisco und Denver flog die Maschine in der Troposphäre. Der Einflug in die Stratosphäre erfolgte wieder nach dem Durchqueren einer Warmluftblase und eines Kältemantels.

Über dem Atlantik wurde die Troposphäre zwischen Neu Fundland und dem 39. Längengrad angeschnitten. Der Temperaturverlauf war auch bei diesen beiden Schnitten ähnlich der beschriebenen. Die stärkste Flugunruhe tritt in dem wärmeren Gebiet auf troposphärischer Seite auf. Die atmosphärischen Erscheinungen waren aber schwächer als über den U.S.A. ; die Maschine wurde über dem Atlantik niedriger und mehr auf der polaren Seite des jetstream geflogen.

Die Erfahrungen in der "cockpit" und die Registrierungen der Flugunruhe deuten daraufhin, daß besonders in der Stratosphäre "clear air turbulence" in sehr dünnen Schichten längs einer Isentropenfläche auftritt. Viele Flugzeugbesatzungen nutzen diese Erfahrungstatsache durch Änderung der Flughöhe und Fahrt, wodurch die statistische Brauchbarkeit der Ergebnisse eingeschränkt wird.

Am 27.3.1963 konnten die Registrierungen des Flugzeuges mit Messungen des Radiosonden-Aufstieges von Flint (72.637) unter sehr günstigen Bedingungen verglichen werden. Der vektorielle Unterschied zwischen beiden Meßverfahren beträgt 0,5 m sec^{-1}.

8. Mikro- und Meso - Zirkulationen an der Tropopause

Bei dem isobaren Schnitt der Tropopause mit dem Flugzeug werden folgende Tatsachen beobachtet und gemessen:

1. Beim Einflug von der Stratosphäre in die Troposphäre folgt hinter der warmen Stratosphären-Luft ein schmales sehr kaltes Gebiet (ungefähr 20 km in Stromrichtung), dann ein warmes Gebiet (ungefähr 20 bis 30 km lang) und erst danach die niedrige Temperatur der Troposphäre. In der Gegenrichtung werden die Verhältnisse entsprechend erflogen.

2. Der erflogene Wind hat in der tiefen Stratosphäre, häufig in der schmalen kalten Schicht ein Maximum.

3. Der erflogene Wind dreht in der kalten Schicht an der Tropopause (Isentropenbuckel) recht, d.h. zur polaren Seite des jetstream.

4. In der wärmeren, tiefer gelegenen Schicht (Isentropengraben) ist ein Rückdrehen des Windes zu bemerken.

5. Im Gebiet der größten Aufsteilung der Isentropen-Flächen (warmer Isentropengraben) herrscht die stärkste, länger anhaltende Flugunruhe (wahrscheinlich durch vertikale Böen ausgelöst).

6. Im Isentropenbuckel an der Tropopause sind die Stöße meist einzeln und kurz (wahrscheinlich durch horizontale Windscherungen ausgelöst).

Die Bilder 27 bis 30 skizzieren schematisch eine mögliche Zirkulation an der Tropopause nach Registrier-Ergebnissen. Die Mikro- und Meso - Zirkulation wird durch dünne Pfeile angedeutet. In allen Darstellungen werden die Isobaren gestrichelt und die Isentropen ausgezogen wiedergegeben.

Bild 27 gibt den Verlauf der mit dem Flugzeug gemessenen Temperatur wieder. Das Flugzeug fliegt isobar und parallel

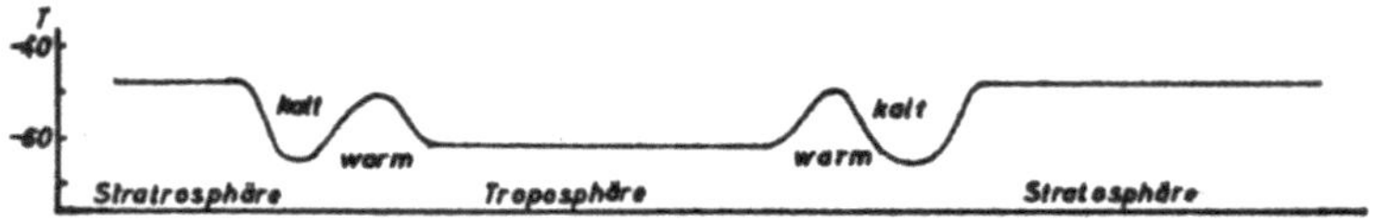

Bild 27: Temperatur im isobaren Flug mit Schnitt der Tropopause

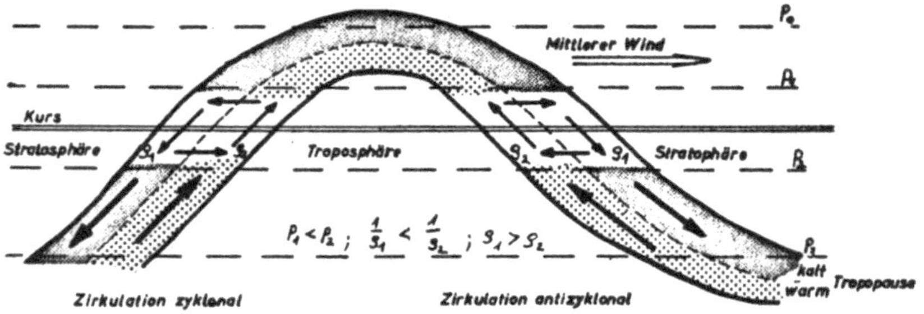

Bild 28: Meso- und Mikro-Zirkulation an der Tropopause Schnitt horizontal

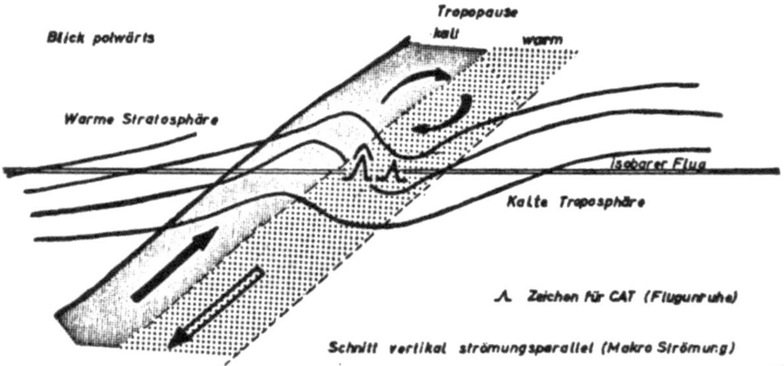

Bild 29: Meso- und Mikro-Zirkulation an der Tropopause

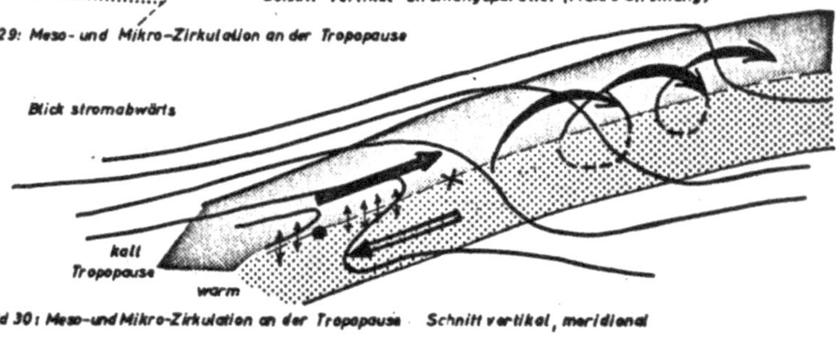

Bild 30: Meso-und Mikro-Zirkulation an der Tropopause Schnitt vertikal, meridional

zur mittleren Strömung von der Stratosphäre in die Troposphäre und wieder in die Stratosphäre.

Bild 28 zeigt einen horizontalen Schnitt durch die Atmosphäre . Der fette Kurvenzug markiert die Schnittlinie der Tropopause . An der troposphärischen Tropopausen-Seite sind die Gebiete mit sehr kalter und warmer Luft unterschiedlich schraffiert. Als Gerade sind der Kurs und die Isobaren eingetragen. Die Isobaren sind gestrichelt und vom tiefen zum hohen Druck mit p_o , p_1 , p_2 , p_3 bezeichnet.

Bild 29 stellt einen vertikalen Schnitt durch die Atmosphäre parallel zur mittleren Strömung und zum Kurs über Grund für den in Bild 28 gezeichneten linken Tropopausen-Durchflug dar. Die Tropopause ist fett markiert. Unterschiedlich schraffierte Gebiete kennzeichnen die sehr kalte Zwischenschicht und die warme Unterschicht unter der Tropopause. Die Gerade bezeichnet das isobare Flugniveau. Die Kurven skizzieren die Schnittlinien der Isentropen-Flächen, wie sie nach den Registrierungen und Vergleichen mit Radiosonden erstellt werden.

Das Bild 30 gibt den vertikalen Schnitt durch die Atmosphäre senkrecht zum mittleren, geostrophischen Wind und zum Kurs über Grund wieder. Die Tropopause ist fett eingetragen. Die Isentropen-Flächen stehen als dünne Linien.

Reiter (51) hebt hervor, daß auf der zyklonalen Seite einer Strahlstrom-Achse ein warmes Gebiet (Isentropengraben) anzutreffen ist, in welchem stratosphärische Luft in die tiefere Troposphäre einbricht. Südlich des jetstream stellt Reiter dagegen eine Kaltluftzunge fest , die er als aufsteigende Luftbewegung deutet . Beide Erscheinungen "Isentropengraben" und "Isentropenbuckel" werden von ihm in ihrer Lage zur Achse des jetstream und seiner Finger betrachtet . Sie sind großräumiger und schließen sich im Bild 29 der gezeichneten Tropopausen-Struktur seitlich an. Der Isentropengraben nach Reiter (besser "Isentropen - Tal" genannt) schließt sich im Bild 29 links auf der polaren stratosphärischen Seite der

Tropopause an. Auf der rechten Seite des Bildes 29 folgt der Isentropenbuckel nach Reiter (besser "Isentropenberg" genannt) dem gezeichneten kleinräumigeren Isentropengraben.

Die Strahlungs- und Strömungsverhältnisse an der Tropopause bestimmen dort die Temperaturen. Die Tropopause kann als Abgleit-Inversion betrachtet werden, die durch das großräumige Abgleiten stratosphärischer Luftmassen von den Tropen zum Polargebiet in der Westwindzone entsteht (Raethjen (41)). Die Tropopause bildet die Obergrenze der Troposphäre und strahlt ungehindert in den Weltenraum ab. Dadurch stellen sich an der Tropopause tiefe Temperaturen ein, welche Luftquanten zum Absinken bringen. Hierdurch könnte eine wärmere Unterschicht entstehen. Unter der Tropopause liegt ein Maximum der Windgeschwindigkeit in der dünnen kalten Oberschicht. Diese Strömung neigt zum Rechtdrehen.

Da die Isentropen-Flächen im planetarischen Mittel von den Subtropen zum Polargebiet abfallen, gleitet die kalte rechtdrehende Strömung auf, die tiefer gelegene warme Strömung gleitet eher ab. Im tiefer gelegenen Niveau kann unter der warmen abgleitenden Luft eine neue Abgleit - Inversion entstehen (Absinken der Tropopause, Blätterigkeit der Tropopause).

An der Tropopause schneiden sich isobare Flächen mit isosteren Flächen. Bild 28 zeigt den wahrscheinlichen Aufbau des Druck- und Dichtefeldes an der Tropopause im horizontalen Schnitt. Die durch das Temperatur- und Dichtefeld entstehenden Strömungen sollen das Druckfeld nicht sofort angleichen. Nach Bild 28 würde auf der linken Seite (zur Strömung aufwärts) eine zyklonale, auf der rechten Seite des Bildes (zur Strömung abwärts) eine antizyklonale Meso-Zirkulation unter der Tropopause entstehen können.

Die Bilder 29 und 30 geben vertikale Schnitte parallel und senkrecht zur mittleren Strömung wieder. Die geschilderten Zirkulationen veranlassen, die Tropopause nicht als Fläche sondern als Schicht zu definieren. Sie besteht aus einer dünnen kalten Oberschicht und einer wärmeren Unterschicht. Ihr Ursprung liegt in der obersten kalten Schicht (Abstrahlungsfläche). Die Tropopause umfaßt die Zirkulation zwischen den beiden genannten Schichten.

Die Ergebnisse der Registrierungen und die Beobachtungen bei Flügen durch die Tropopause etwa parallel zur geostrophischen Strömung können durch folgendes Zirkulationsmodell erklärt werden:

> Die in der obersten Tropopausen-Schicht entstehende kältere Luft fällt in der mittleren Strömung stromabwärts. Sie erwärmt sich adiabatisch durch Absinken und strömt annähernd isentrop in die Richtung der polaren Seite des jetstream. Die entstehende wärmere Luft neigt zu Aufsteigen gegen die mittlere überregionale Strömung (Bilder 28 bis 30), falls die Oberschicht kälter wird (vertikale Labilisierung).

Dieser Vorgang führt zu schraubenartigen (Flecht-) Strömungen (Prandtl (39)). Hierdurch entstehen straßenförmige Anordnungen der Turbulenz-Gebiete sowie Cirrus-Banden. Derartige Wolkenstrukturen im Bereich des jetstream sind durch Kuttner (33) beschrieben. Bandenförmige Wolkenformen an niedrigen Inversionsschichten behandelte Seilkopf (61).

Die anscheinend turbulenten Bewegungen in Wolken und an Inversionen können geordneten, kleinräumigen Zirkulationssystemen zugeschrieben werden. Die Erkenntnis, daß an der Tropopause geordnete Strömungen herrschen, zwingt den Begriff "Turbulenz" für die dortigen Vorgänge einzuschränken. Als Turbulenz bezeichnet man eine stark wirbelnde Durchmischung der Grenzschicht zwischen zwei Strömungen mit verschiedenen Geschwindigkeiten. Bei der Turbulenz ballen sich die Strömungselemente im ungeordneten Austausch zu ständig wechselnden Gebilden zusammen. Der Begriff sollte in der Meteorologie dort verwendet werden, wo ungeordnete Prozesse erkennbar sind. Die kleinräumigen Zirkulationen an der Tropopause erscheinen nur in der

Makrostruktur als turbulent. In der Mesostruktur ist eine gewisse Ordnung erkennbar. Die Makrostruktur gilt als geordnet. Die Erkenntnis über die räumliche Ausdehnung geordneter Prozesse ist auch hier von dem Auflösungsvermögen der gewählten Meßapparatur abhängig.

9. Zur Methodik und Genauigkeit der Messung erflogener Winde

9.1 Meßträger, Instrumente, Fehler

Der Meßträger (Boeing 707 - 430 D-ABOD) für das Registriergerät ist mit einem Doppler-Navigationssystem ausgerüstet. Das Doppler-Radar-Gerät nutzt zur Messung der Drift die Doppler-Effekte von drei Radarstrahlen (zwei schräg voraus, einer schräg nach rückwärts gesendet). Zusätzlich ist in die Maschine zur genauen Temperatur - Registrierung ein Rosemount - Thermometer und ein Vertikalbeschleunigungsgeber eingebaut. Das Rosemount - Thermometer mißt außerhalb der Reibungsschicht des Flugzeuges (vorn an der rechten Außenseite unterhalb der Flugzeugkanzel) eine Temperatur, deren adiabatische Erhöhung im Flug definiert ist. Schräg unterhalb des Rosemount - Thermometers befindet sich das Pitotrohr des Flugzeuges zur Bestimmung des Staudruckes. Die Abnahme des statischen Druckes erfolgt auf beiden Seiten des Flugzeugrumpfes etwa zwei Meter vor der Tragflächen-Vorderkante am unteren Drittel des Rumpfes. Das Registriergerät und der Vertikalbeschleunigungsgeber mit sämtlichen Zusatzgeräten befinden sich im Frachtraum (Schwerpunkt) des Flugzeuges.

Der eingebaute SFIM-Vielfachschreiber mit der Typenbezeichnung A 2610 zeichnet die Daten über eine Lichtspur auf photographischem Papier nieder. Die Registrierung erfolgt auf Schreibstreifen von 25 m Länge mit einem Vorschub von 17 mm/min. Die routinemäßig eingebauten Fluginstrumente geben die Signale über Umformersätze in digitaler Schaltung auf das Registriergerät. Für den Temperatur-Meßkreis des Rosemount-Thermometers wird mit Analog-Schaltung gearbeitet. Im Schreibgerät gelangen die elektrischen Signale über Quotienten-Meßwerke zur Anzeige. Das Registriergerät besitzt Druckmeßwerke für den statischen Druck und den Staudruck.

Das Gerät wurde im Labor und am Flugzeug im eingebauten Zustand pneumatisch und elektrisch geprüft. Im Rahmen dieser Untersuchung stimmen die Instrumenten-Ablesungen, welche durch die Flugbesatzungen in die Navigationslogs eingetragen werden, mit den Anzeigen auf den Schreibstreifen gut überein. Drei Versuche wurden unternommen, durch einen zusätzlichen Beobachter die Fluginstrumente während des Fluges laufend zu überwachen. Es zeigen sich keine systematischen Fehler.

Nach Radiosonden-Messungen wurden zu diesen Flügen durch die Atmosphäre Schnitte gelegt, die parallel und senkrecht zum Kurs über Grund verlaufen. Die Ergebnisse aus den Schnitten stimmen sinnvoll mit den Registrierungen überein. Es gelang ein direkter Vergleich mit einem amerikanischen Radiosonden-Aufstieg (siehe Seiten 80, 124). Der vektorielle Unterschied der Winde nach den beiden Meßverfahren beträgt 0,5 m sec^{-1}.

Die "systematischen Fehler" der Registrieranlage sind aber nur dann vernachlässigbar klein, wenn die systematischen Fehler und Eichkurven berücksichtigt werden, die dem von der Federal Aviation Agency (FAA) geprüften Flight-Manual für Boeing 707 (Boeing (5)) zu entnehmen sind. Die dortigen Werte ergeben sich aus Testflügen mit Schleppsonden und Vergleichsflügen. Boeing verzichtet auf eine besondere Eichung der einzelnen Meßeinrichtungen in den Serien-Flugzeugen durch Eichflüge und Vergleichen mit anderen Flugzeugen, weil die Herstellungstoleranzen von FAA niedrig eingeschätzt werden.

Der Gesamtfehler der Registrieranlage ergibt sich nach den Gesetzen der Fehlerfortpflanzung aus den Beiträgen der Fehler in den Meßelementen, den Übertragungs- und Umformungseinrichtungen, sowie in dem Registriergerät. Daneben sind bei der Ablesung und Auswertung weitere Ungenauigkeiten zu berücksichtigen, die aber unbedeutend gegenüber den vorher genannten sind. Die Tabelle 7 stellt die registrierten Elemente mit den gewählten Auflösungen und den mittleren Fehlern (Angaben der Boeing-Company und der Deutschen Lufthansa für Ausgangswerte, eigene Untersuchungen für abgeleitete Größen) zusammen.

Tabelle 7: Auflösung und mittlere Fehler der registrierten Elemente auf den Schreibstreifen

Keine Ausgleichung der registrierten Kurven auf Schreibstreifen

Element		Auflösung pro mm	Mittlerer Fehler
Temperatur	grob	0,9 °	2 °
	fein	0,1 °	0,01 °
Grundgeschwindigkeit		3,1 m sec^{-1}	2,2 m sec^{-1}
Driftwinkel		0,7 °	0,7 °
Steuerkurs		7 °	1 °
Staudruck		2,7 m sec^{-1}	1,0 m sec^{-1}
Druckhöhe	fein	300 ft	100 ft
Vertikale Beschleunigung			
26.3.1962 bis 31. 5.1962		80 cm sec^{-2}	60 cm sec^{-2}
1.1.1963 bis 31.10.1963		40 cm sec^{-2}	30 cm sec^{-2}
Berechnete Größen:			
Flugzeugeigengeschwindigkeit			1,4 m sec^{-1}
erflogene Windgeschwindigkeit			2,9 m sec^{-1}
erflogene Windrichtung bei		5 m sec^{-1} Wind	32°
		15 m sec^{-1} Wind	11°
		30 m sec^{-1} Wind	5°
		50 m sec^{-1} Wind	3°

Die Methodik zur Berechnung des mittleren Fehlers für den erflogenen Wind wird in den Abschnitten 9.2 und 9.3 skizziert. In allen Zweifelsfällen bei der Größe der mittleren instrumentellen Fehler wurde stets zur ungünstigen Seite abgeschätzt. Die in Tabelle 7 genannten Größen für die mittleren Fehler der registrierten Elemente werden niedriger, falls eine nachträgliche Ausgleichung der Kurven stattfindet. Einstellzeiten, Dämpfung, Empfindlichkeit und Auflösung der einzelnen Instrumente und Signale sind nicht gleich. Die registrierten Parameter liefern zum mittleren Fehler des berechneten Windes dadurch höhere Beiträge als bei guter Anpassung.

Durch günstige nachträgliche Ausgleichung und Abstimmung der Kurven für Grundgeschwindigkeit, angezeigte Flugzeugeigengeschwindigkeit, Driftwinkel und Steuerkurs vermindern sich die mittleren Fehler des Windes auf die in Tabelle 8 angegebenen Größen. Diese Werte werden von den über die KIFIS-Anlage im Flugzeug errechneten "spot-winds" nicht erreicht.

Tabelle 8: Mittlerer Fehler des erflogenen Windes bei nachträglicher Ausgleichung der registrierten Kurven (Bezeichnungen nach Bild 31 , Seite 106)

	Wind	Mittlere Fehler	
für Windwinkel ω =	—	0°	90°
erflogene Windgeschwindigkeit	—	1,4 m sec^{-1}	2,0 m sec^{-1}
erflogene Windrichtung bei	5 m sec^{-1}	24°	16°
	10	12°	8°
	15	8°	6°
	20	6°	4°
	30	4°	3°
	40	3°	2°
	90	1°	1°

Die Tabelle 8 wurde mit folgenden mittleren instrumentellen Fehlern errechnet:

Flugzeugeigengeschwindigkeit m_v = 1,0 m sec^{-1}

Grundgeschwindigkeit m_u = 1,0 m sec^{-1}

Driftwinkel m_a = 0,7 °

Sußebach (65) bezweifelt die oben genannten Fehlergrößen und glaubt, daß für kleine Windgeschwindigkeiten die Windrichtung sehr viel genauer angegeben werden kann (hierzu sei auf den Abschnitt 9.3 "Mittlerer Fehler des erflogenen Windvektors bei Benutzung des Doppler-Radar-Gerätes" verwiesen).

Die im Flugzeug empfangene Doppler-Frequenz eines Radarstrahles ist nur von der Geschwindigkeitskomponente abhängig, mit der sich das Flugzeug relativ zu der augenblicklich angestrahlten und reflektierenden (besser: streuenden) Fläche bewegt. Sofern die reflektierende Materie ortsfest ist, entstehen keine

Verfälschungen (auch nicht durch Neigungen des Untergrundes, wie Berge, Meereswellen). Eine bewegte Meeresoberfläche (wie Gischt, Wasserfetzen) kann aber beträchtliche Fehler im Grundgeschwindigkeitsvektor hervorrufen (bei Orkan in Meereshöhe über 3 m sec^{-1} falsche Anzeige).

Wolken verfälschen die Meßergebnisse praktisch nicht, weil auf dem verhältnismäßig steilen Strahlenweg vom Doppler-Radar-Gerät zum Boden die Wolken vom Radarstrahl gut durchdrungen werden und die im Flugzeug empfangene Energie von Wolken gegenüber derjenigen vom Boden unbedeutend ist.

Da der Empfänger im Flugzeug durch die Neigung des Radarstrahls nur gestreute (nicht reflektierte) Energie zur Messung nutzen kann, wird über glatten Meeresgebieten weniger Energie aus flach angestrahlten Gebieten, die höhere Frequenzverschiebungen erzeugen, als aus steil angestrahlten Gebieten, die niedrigere Frequenzverschiebungen erzeugen, aufgenommen. Hierdurch entsteht im empfangenen Spektrum eine Schwerpunktsverschiebung, die über See eine um ungefähr 1,5% niedrigere Anzeige der Grundgeschwindigkeit bewirkt (Ausgleich über Schalterbetätigung "Land-See" am Doppler-Radar-Gerät im Flugzeug vorgesehen).

Eine weitere Verfälschung des Windes kann durch die hohe Einstellzeit der Grundgeschwindigkeit gegenüber derjenigen der Flugzeugeigengeschwindigkeit entstehen. Die "Nachstellzeit" beträgt bei einer Auslenkung (Ablage) der Grundgeschwindigkeitsanzeige von 50 m sec^{-1} bis zum Wert 0 ungefähr 130 sec (der Wert 5 m sec^{-1} ist nach 65 sec erreicht). Da die Registrierungen der Flugzeugeigengeschwindigkeit entsprechend ausgeglichen wurden, sind für die in der vorliegenden Arbeit benutzten Windwerte keine höheren Fehlerbeiträge als die in Tabelle 8 angegebenen Werte zu berücksichtigen. Es sei zur Genauigkeit der an die Flugwetterwarten übermittelten "spot-winds" vermerkt, daß höhere Fehler in der Windrichtung zusätzlich durch Fehler im erdmagnetischen Steuerkurs entstehen können.

9.2 Windbestimmung mit dem Flugzeug

Der erflogene Wind wird als die vektorielle Differenz zwischen der Grundgeschwindigkeit und der Flugzeugeigengeschwindigkeit definiert. Er wird aus der Versetzung des Flugzeugs bestimmt:

1. Die Versetzung wird navigatorisch aus zwei Standortsbestimmungen ermittelt.
2. Die Versetzung wird mit Hilfe des Doppler-Radar-Gerätes gemessen.

Bild 31 zeigt die Bezeichnungen im "Winddreieck". Zu der Flugzeugeigengeschwindigkeit (v) , die in die Richtung der Längsachse des Flugzeuges weist und durch den Schub der Turbinen bestimmt wird, addiert sich die Windgeschwindigkeit (w) und ergibt die Geschwindigkeit des Flugzeuges über Grund ($\breve{u}$), kurz Grundgeschwindigkeit genannt. Zwischen der Richtung der Flugzeugeigengeschwindigkeit (K) , genannt Steuerkurs, und der Richtung der Grundgeschwindigkeit (ε) , Kurs über Grund genannt, liegt der Driftwinkel (α).

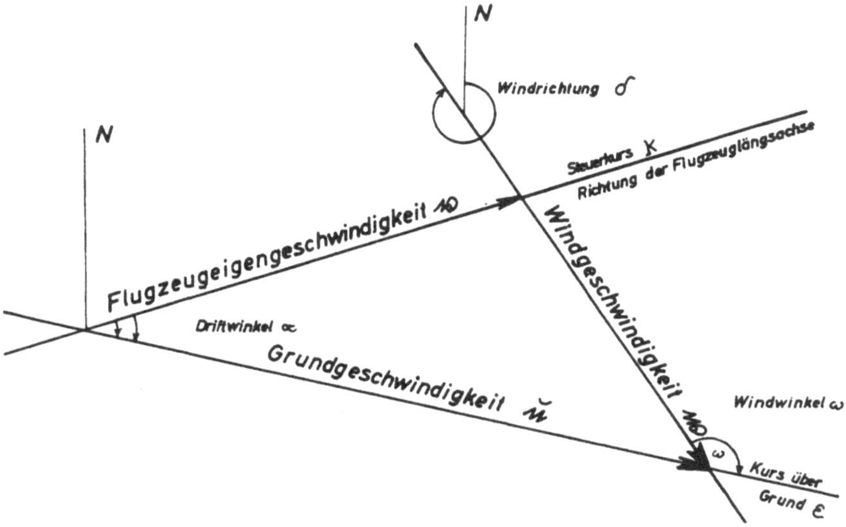

Bild 31: Winddreieck

Die Problematik einer Windbestimmung mit dem Flugzeug offenbart sich in den beiden erwähnten Methoden. Ein Wind, der aus der Versetzung des Flugkörpers über eine durchflogene Meßstrecke berechnet wird, ist stets ein "mittlerer Wind". Dieser Wind wird auf den Mittelpunkt der Meßstrecke bezogen und widerspricht in all jenen Fällen der Wetterkarte, wo nichtlineare Änderungen der Windgeschwindigkeit oder größere Windrichtungsänderungen auftreten (z.B. Flug durch Trog mit asymetrischen Windfeld). Auch ein "spot-wind", der aus den Werten der Doppler-Radar-Anlage und dem KIFIS für einen Ort der Flugstrecke bestimmt wird, ist ein "integrierter Wind". Die Integration erfolgt über eine Flugstrecke, deren Länge durch die Einstellzeiten der Fluginstrumente bestimmt wird. Bei kurzen Integrationsstrecken zeigt ein "spot-wind" starke Abweichungen von der Wetterkarte. Die Wetterkarte stellt einen aus Radiosonden-Messungen gemittelten Zustand der Atmosphäre zu einem Zeitpunkt dar und berücksichtigt nicht kleinräumige Unterschiede im Temperatur- oder Windfeld. Das Doppler-Radar erfaßt kleinräumige Windunterschiede.

Es sind die Forderungen gegeneinander abzuwägen:

1. Der erflogene Wind soll gut in das Bild einer Wetterkarte passen (der heutige Stand in der Grob-Analyse über den Ozeanen sollte nicht Maßstab sein).
2. Der erflogene Wind muß für die Vorhersage des Gradientwindfeldes gut geeignet sein; er muß die Makrostruktur der Atmosphäre genau wiedergeben.
3. Wetterhafte Erscheinungen sollten im Windfeld erfaßt werden. Die Mesostruktur muß aufgenommen werden.
4. Der erflogene Wind darf nicht durch Flugeigenschaften verfälscht werden. Die Mikrostruktur muß weitgehend unterdrückt werden.

Die Forderung 1 ist für den Atlantik heute erfüllt, wenn die erflogenen Winde über eine Strecke von ungefähr 600 km geglättet werden (Wetterkarte nur mit Radiosonden-Werten). Die Forderung 3 dürfte bei einer Mittelung des erflogenen Windes über 200 bis 300 km ausreichend erfüllt sein. Die Forderung 4 ist befriedigt, sofern die registrierten Parameter über eine Flugstrecke von einer Minute ausgeglichen werden. Die bei der Doppler-Radar-Anlage eingeführte Empfindlichkeit, Dämpfung und Einstellzeit unterdrücken Verfälschungen durch Flugeigenschaften in turbulenter Luft. Die Einflüsse der Mikrostruktur auf angezeigte Wind-Werte wären beseitigt, falls die Parameter der Flugzeugeigengeschwindigkeit in gleicher Weise behandelt würden. Ein sehr günstiges Bild für die Analyse einer Wetterkarte liefern erflogene Winde, falls sie über eine Flugstrecke von sechs Minuten (ungefähr 90 km) integriert werden.

Ein derartiges Windmittel ist nicht notwendig auch für die Zukunft das beste. Das Wetterkartenbild stellt heute ein statistisches Mittel einer vorgegebenen Größenklasse dar, die durch den Abstand der Radiosonden-Stationen bestimmt ist. Die Zukunft wird lehren, ob mit Hilfe der elektronischen Datenverarbeitungsmaschinen besser Mittelwerte mit Streuung oder eine große Zahl von Einzelwerten für die Analyse einer Wetterlage verwendet werden.

Die mitgeteilten Ergebnisse gründen auf Erfahrungen und Versuchen, die durch verschiedene Ablesungen und Integrationen der registrierten Kurven gewonnen wurden. Da die Registrierungen nur durch ein Düsenflugzeug vom Typ Boeing 707-430 erflogen wurden, sind die Ergebnisse streng nur für diesen Flugzeugtyp mit der installierten Einrichtung gültig.

9.3 Mittlerer Fehler des erflogenen Windvektors bei Benutzung des Doppler-Radar-Gerätes

Die Grundgeschwindigkeit und der Driftwinkel sind Funktionen des Windes. Sie liegen im Flugzeug als instrumentell gegebene Größen des Doppler - Radar - Gerätes vor. Die Flugzeugeigengeschwindigkeit ist im Flugzeug ebenfalls zu bestimmen. Die Differenz aus den beiden Geschwindigkeiten wird hier als "erflogener Wind" bezeichnet. Dabei handelt es sich nicht nur um einen mittleren Wind, der aus der Versetzung des Flugzeuges berechnet wird, sondern auch um Beiträge zu dem Windvektor, die durch systematische und zufällige Navigationsfehler oder Fehler in der Anlage hinzukommen. Erflogene Winde sind vom Flugzeug-Typ abhängig, sofern die Flugeigenschaften sich in Gebieten mit Flugunruhe (CAT) unterscheiden und Einfluß auf die Meßwerte bekommen.

Zu allen Fehlerberechnungen wird das Fehlerfortpflanzungsgesetz von Gauss benutzt. Es beantwortet die Frage, in welcher Weise der mittlere Fehler einer errechneten Größe von den mittleren Fehlern der Teilgrößen abhängt, aus denen sich die Größe zusammensetzt. Dabei wird vorausgesetzt, daß die Meßfehler der Ausgangsgrößen relativ klein gegen die Größe selbst, der "unabhängig Veränderlichen" sind.

Die mittleren Fehler der registrierten Elemente sind allgemein klein gegen den Meßwert. Bei geringen seitlichen Versetzungen ist die Größe des Driftwinkels klein, aber die Fehler nicht notwendig klein. Obwohl die Voraussetzungen zum Fehlerfortpflanzungsgesetz nur eingeschränkt für den abzuleitenden mittleren Fehler des Windvektors zutreffen, sollen mit dem Fehlerfortpflanzungsgesetz die mittleren Fehlergrößen abgeschätzt werden. Der Einzelmessung kann bei geringen Windgeschwindigkeiten nicht viel Vertrauen geschenkt werden, weil die Meßfehler von der Größenordnung der Meßwerte sind. Zum allgemeinen Verständnis sei wiederholt, daß die Grundgeschwindigkeit und der Driftwinkel in der Fehlerberechnung als unabhängige Veränderliche, der Wind dagegen als abhängige Veränderliche betrachtet wird. Der Vektor der Grundgeschwindigkeit ist im Flugzeug meßbar.

Die Gleichung für den mittleren Fehler des Windvektors ist in der unveröffentlichten Arbeit: "Die Genauigkeit der Windbeobachtungen, welche von Verkehrsflugzeugen der Deutschen Lufthansa in den Monaten Oktober 1959 bis März 1960 als Abdrift erflogen wurden" abgeleitet (13). Der folgende Text gibt nur die wesentlichen Gedanken wieder.

Das mittlere Fehlerquadrat der Komponente des Windvektors, die senkrecht zum Kurs über Grund liegt, lautet:

$$(9.3.1) \qquad m_y^2 = \sin^2\alpha \cdot m_v^2 + v^2 \cos^2\alpha \cdot m_\alpha^2$$

Das mittlere Fehlerquadrat der Komponente des Windvektors, die parallel zum Kurs über Grund liegt, lautet:

$$(9.3.2) \qquad m_x^2 = m_u^2 + \cos^2\alpha \cdot m_v^2 + v^2 \cdot \sin^2\alpha \cdot m_\alpha^2$$

m_u mittlerer Fehler der Grundgeschwindigkeit u
m_v mittlerer Fehler der Flugzeugeigengeschwindigkeit v
m_α mittlerer Fehler des Driftwinkels α

Die Größen sind in Bild 31 (Seite 103) erklärt.

Der Betrag des Windvektors gibt die Windgeschwindigkeit an:

$$(9.3.3) \qquad |\mathbf{w}| = w = \sqrt{u^2 + v^2 + 2 \cdot u \cdot v \cdot \cos\alpha}$$

Der mittlere Fehler der Windgeschwindigkeit lautet:

$$(9.3.4) \qquad m_w = \frac{1}{w}\sqrt{(u - v\cdot\cos\alpha)^2 \cdot m_u^2 + (v - u\cdot\cos\alpha)^2 \cdot m_v^2 + (u\cdot v\cdot\sin\alpha)^2 \cdot m_\alpha^2}$$

Die Abkürzung w steht nach Gleichung (9.3.3).
Für die Werte w , u , v und α sind die Mittelwerte der zu untersuchenden Kollektive einzusetzen.

Nach Gleichung (9.3.4) ist eine Abhängigkeit des mittleren Fehlers der Windgeschwindigkeit vom Wind zu erwarten. Je nach Lage des Windvektors zu den Vektoren der Flugzeugeigengeschwindigkeit und der Grundgeschwindigkeit wirken sich die Fehlerbeiträge auf den Gesamtfehler verschieden stark aus. Die angeführten mittleren instrumentellen Fehler sind unabhängig vom Wind. Aus praktischen Erwägungen werden die Größen u, v und α in Gleichung (9.3.4) mit Hilfe trigonometrischer Beziehungen im Winddreieck durch den Windwinkel ω und die Windgeschwindigkeit w formal ersetzt. Unter der Voraussetzung, daß die Flugzeugeigengeschwindigkeit sehr viel größer als die Windgeschwindigkeit ist, folgt für den mittleren Fehler der Windgeschwindigkeit vereinfacht:

$$(9.3.5) \quad m_w = \sqrt{(\cos\omega)^2 \cdot m_u^2 + (\cos\omega)^2 \cdot m_v^2 + (v \cdot \sin\omega)^2 \cdot m_\alpha^2}$$

Mit x_w und y_w als Koordinaten des Windvektors **w** (siehe wie Gleichungen (9.3.1) und (9.3.2)) ist das Fehlerquadrat des Windwinkels ω durch:

$$(9.3.6) \quad m_\omega^2 = \left(\frac{\partial\omega}{\partial x_w}\right)^2 \cdot m_{x_w}^2 + \left(\frac{\partial\omega}{\partial y_w}\right)^2 \cdot m_{y_w}^2$$

gegeben.

Wir ersetzen mit Hilfe der trigonometrischen Beziehungen im Winddreieck die Grundgeschwindigkeit und den Driftwinkel durch die Windgeschwindigkeit und den Windwinkel. Es folgt vereinfacht für den Fall, daß die Windgeschwindigkeit klein gegen die Flugzeugeigengeschwindigkeit ist:

$$(9.3.7) \quad m_\omega = \sqrt{\frac{1}{w^2}\left(\sin^2\omega \cdot m_u^2 + \sin^2\omega \cdot m_v^2 + (w^2 \cdot \sin^4\omega + v^2 \cdot \cos^2\omega)\, m_\alpha^2\right)}$$

Der mittlere Fehler der Windrichtung ist sehr wesentlich von der Windgeschwindigkeit abhängig. Er wird um so größer, je kleiner die Windgeschwindigkeit ist. Bei sehr kleinen Windgeschwindigkeiten liegen die Spitzen der vom Flugzeug meßbaren Vektoren (Flugzeugeigengeschwindigkeit w , Grundgeschwindigkeit u) so nah beieinander und die Fehler der beiden Vektoren sind so groß, daß eine Windrichtung nicht mehr sinnvoll angegeben werden kann. Die nach Gleichungen (9.3.5) und (9.3.7) berechneten mittleren Fehler des erflogenen Windes sind in Tabelle 7 und Tabelle 8 angegeben und wurden im Abschnitt 9.1 auf den Seiten 103 und 104 besprochen.

Sußebach (65) bezweifelt die in dieser Arbeit angegebenen Fehlergrößen, deren Ableitungen und Werte den **unveröffentlichten** Arbeiten des Verfassers (13,14) entnommen sind. Sußebach kritisiert die Fehlerangaben besonders für kleine Windgeschwindigkeiten, weil während des Fluges durch einen Höhentrog mit starker Drehung des Windes bei kleinen Windstärken die registrierten Werte nur wenig streuen und keiner als ausgesprochen falsch anzusehen sei. Hierzu muß klargestellt werden, daß der erflogene Wind nur aus der Verdriftung des Flugzeuges bestimmt wird. Bei kleinen Windgeschwindigkeiten ist die Meßgenauigkeit nicht ausreichend, um eine geringe Verdriftung überhaupt feststellen zu können. Außerdem ist es sinnlos, für Windstille eine Windrichtung anzugeben (der mittlere Fehler der Windrichtung ist für Windstille unabhängig vom Meßverfahren stets unendlich groß). Eine große Streuung der erflogenen Windrichtung gegenüber der tatsächlich herrschenden Strömung läßt sich aber nur in ausgedehnten windstillen Gebieten (z.B. Hochdruckgebieten) experimentell bestätigen. Die Vektoren Grundgeschwindigkeit und Flugzeugeigengeschwindigkeit sind dort über längere Zeiten ungefähr gleich groß (aber nur innerhalb der Meßgenauigkeit, es kann also auch ein geringer Wind vorgetäuscht werden). Für einen Flug durch einen schmalen Trog liegen zur Berechnung der höheren Windgeschwindigkeiten auf der Trog-Vorderseite und auf der Trog-Rückseite eindeutig meßbare größere Verdriftungen vor. In der Trogachse dagegen nähern sich einander während des Durchfluges die Vektoren Grundgeschwindigkeit (etwas verzögert durch hohe Einstellzeit) und Flugzeugeigengeschwindigkeit durch Änderung des Windes. Dabei wird durch die stetige Abnahme oder Zunahme des Grundgeschwindigkeitsvektors eine hohe Genauigkeit der Windrichtung für kleine Windgeschwindigkeiten vorgetäuscht. Die in den Tabellen 7 und 8 angegebenen Werte wurden nach elektronischen Berechnungen erstellt, wobei keine Glieder der weiter entwickelten Gleichungen (9.3.4) und (9.3.6) vernachlässigt wurden. Die theoretischen Fehlerwerte variieren je nach Lage des Windvektors zu den Fluggeschwindigkeitsvektoren, wie aus Tabelle 8 zu ersehen.

9.4 Einflüsse auf das Verhalten der registrierten Parameter durch das Benehmen des Meßträgers bei verschiedenen atmosphärischen und flugbetrieblichen Bedingungen

9.4.1 Schwingungsformen des Meßträgers

Die Mikrostruktur der Luftströmungen wird auch durch Transosonden und Rauchabsonderungen aus Raketen untersucht. Die Bewegung dieser nicht-starren Meßkörper (Ballon, Rauchmenge), die für ein bestimmtes Flugniveau ausgewogen sind, ist schwierig zu verfolgen. Die Körper werden durch vertikale Impulse zu Schwingungen um ihr Gleichgewichtsniveau angeregt, so daß die tatsächlichen vertikalen Luftbewegungen nicht zu bestimmen sind.

Langperiodische und kurzperiodische Schwingungen überlagern auch die Flugbewegungen des Meßträgers vom Typ Boeing 707. Die langperiodischen Schwingungen (Phygoid) sind zwar schlecht gedämpft aber aussteuerbar. Ihre Amplituden sind allgemein gering. Sie sind zeitweilig in der Registrierung der Druckhöhe und des Steuerkurses zu erkennen (etwa 0,01 Hz und kleiner). Durch den Ausfall der automatischen Steuerungsanlage (Autopilot) wurde auf einem Flug eine Schwingung größerer Amplitude angeregt. Die festgestellten Werte sind umseitig zusammengestellt. In diesem Fall lag eine Schwingung geringer Frequenz (ungefähr eine Schwingung pro Minute) mit großer Amplitude vor. Die Dämpfung der Doppler-Radar-Anlage sowie die Trägheit reichen in derartigen Fällen nicht aus, um die starken Auslenkungen der Radarstrahlen auszugleichen.

Tabelle 9: Schwingungen der registrierten Parameter nach Ausfall der automatischen Steuerungsanlage

Element	Amplitude	
Statischer Druck	150	ft
Staudruck	2	$m\ sec^{-1}$
Grundgeschwindigkeit	1,5	$m\ sec^{-1}$
Driftwinkel	0,5	°
Steuerkurs	2	°
Vertikale Beschleunigung	1,5	$m\ sec^{-2}$
Temperatur (KIFIS)	0,7	°
Temperatur (Rosemount)	ausgefallen	
Frequenz	0,015	Hz
Wind	Berechnung unmöglich	

Während eines Fluges zwischen Frankfurt und San Francisco wurden die normal auftretenden Schwingungen des Flugzeuges grob überprüft. Neben einer deutlich wahrzunehmenden Schwingung von ungefähr sechs Sekunden Dauer sind verschiedene Schwingungen von mehreren Minuten Dauer festzustellen. Es bleibt ungewiß, inwieweit hierbei atmosphärische Einflüsse mikrostruktureller Art wesentlich mitspielen oder Flugeigenschaften des Meßträgers in Erscheinung treten. Die Berechnung des Windes wird durch diese Schwingungen mit geringer Amplitude nicht beeinträchtigt.

"Clear Air Turbulence" regt sehr viel höhere Frequenzen der Schwingungen in der Flugbewegung an. Diese kurzperiodischen Schwingungen wirken sich auch auf die Registrierungen des Driftwinkels und der Grundgeschwindigkeit aus. Die Rechenanlage des Doppler-Radar-Systems schaltet automatisch bei stärkeren kurzperiodischen Schwingungen auf " memory ", d.h. es wird der letzte eindeutig erfaßte Radarreflex gespeichert und in ausgewerteter Form angezeigt. Eine derartige Speicherung kann auch über ruhigem Wasser, über Wüstengebieten und bei Kurven- und Steigflügen eintreten. Die Berechnung des Windes ist für diese Fälle unmöglich.

9.4.2 Massenverlust des Meßträgers im Langstreckenflug

Eine Boeing 707 verbraucht während eines Fluges über den Nordatlantik an Brennstoff etwa ein Drittel ihres Gesamtgewichtes. Der erhebliche Massenverlust wirkt sich auf das Flugverhalten des Meßträgers aus. Durch die Gewichtsabnahme verändert sich der Anstellwinkel der Tragflächen gegenüber der anströmenden Luft. Es könnten Verfälschungen in den Temperatur-, Staudruck-, statischen Druck- und Vertikalbeschleunigungsanzeigen eintreten. Das Rosemount-Thermometer ist nach Angaben des Herstellerwerkes nur wenig neigungsempfindlich. Die mittlere Lage der Vertikalbeschleunigungsanzeige verändert sich während eines Fluges über den nördlichen Atlantik bis zu 1 mm auf dem Schreibstreifen. Das Turbulenz-Spektrum könnte nicht einheitlich registriert werden, weil das Flugzeug durch den Verlust an Masse verschieden träge ist. Die Neigungsempfindlichkeiten der Druckanzeigen, die wesentlichen Einfluß auf die Größe des Windes hätte, soll unbedeutend sein, konnte aber nicht direkt von der Forschungsgruppe bei der Deutschen Lufthansa untersucht werden. Es wurden mit Hilfe der elektronischen Rechenmaschine der Universität Hamburg Unterschiede zwischen den Windfeldern der ersten und letzten Stunden eines Fluges gesucht.

In einem Flugzeug-bezogenen Koordinatensystem weichen die Querwindkomponente (hier: normal zur Rumpfachse), die Längswindkomponente (hier: parallel zur Rumpfachse) und der Betrag des Windvektors, sowie auch seine Änderung (vektorielle Wind-Differenz in zeitlichem Abstand von sechs Minuten Flugzeit) zwischen den ersten und letzten Flugstunden nicht überzufällig voneinander ab. Die Korrelationsfaktoren, welche sich bei Vergleichen der ersten und letzten zehn Werte eines Fluges ergeben, liegen im Zufälligkeitsbereich. Der Massenverlust des Meßträgers kann folglich den "erflogenen Wind" nicht wesentlich verfälschen.

9.4.3 Unterschiede in den Meßergebnissen aus verschiedenen Registrierintervallen

Zur Überprüfung der Windberechnungen wurden die Daten nach vier Zeitabschnitten unterschieden:

Registrier-Periode	1962	März	-	Mai
Registrier-Periode	1963 - I	März	-	Juni
Registrier-Periode	1963 - II	Juli		
Registrier-Periode	1963 - III	August	-	Oktober

Die Daten wurden nach Flugstunden sortiert und in folgenden Koordinatensystemen untersucht:

1. geographisch, Nord und Ost;
2. flugzeug-bezogen, parallel und normal zur Rumpfachse;
3. strömungsbezogen, parallel und normal zu der herrschenden atmosphärischen Strömung.

Die mittlere Streuung war in den Kollektiven der dritten und vierten Registrier-Periode höher als in den restlichen. Auffällig waren hohe Werte in den ersten und letzten Flugstunden, gegenüber niedrigeren Streuungen für den mittleren Teil der Atlantik-Überquerungen. Bei kritischer Durchsicht der Einzelwerte stellte sich heraus, daß die Windberechnungen in den letzten Monaten der Registrier-Periode 1963 nach anderen Anweisungen bei der Deutschen Lufthansa vorgenommen sein mußten als in der vorhergehenden Zeit.

Die Aufbereitungsgruppe bei der Deutschen Lufthansa arbeitete aus finanziellen Gründen ab August 1963 ohne wissenschaftliche Leitung. In dieser Zeit wurden Windberechnungen auch aus Daten vorgenommen, die während oder kurz nach Kurs- oder Höhenänderungen registriert wurden. Nachdem die fehlerhaften Werte selektiert waren , ergibt die Rechnung in allen Intervallen eine gleichmäßige Streuung des Windes und der Winddifferenzen zwischen zwei Meßpunkten. Die Ergebnisse der vorangehenden Abschnitte wurden mit diesen nachträglich verbesserten Daten erzielt.

9.4.4 Streckenplanung und flugbetriebliche Maßnahmen

Die Flugzeuge fliegen zwischen den Kontinenten auf dem Wege kürzester Flugzeit (Minimum Time Track). Es wird der Flug, welcher ohne Wind längs des Großkreises erfolgen würde, so abgewandelt, daß die günstigsten Winde genutzt werden. Auf dem Flug von Amerika nach Europa wird meist im Gebiet der größten Westwindgeschwindigkeit geflogen. Der Flug von Europa nach Amerika liegt meist nördlicher. Er weicht den Westwinden aus und erfolgt überwiegend auf der polaren Seite des Strahlstromes.

Durch diese Flugplanung werden geographische Räume nur zu speziellen Wetterlagen überflogen. Diese Auswahl mag aus wissenschaftlicher, besonders klimatologischer Sicht ungünstig sein, sie bietet jedoch gerade für die zivile Luftfahrt Aufschlüsse aus den für sie interessanten Gebieten. Eine Betrachtung meteorologische Phänomene nach geographischen oder klimatologischen Gesichtspunkten scheidet aus. Dafür bietet sich an, die Daten nach dem jeweiligen Strömungsbild zu orientieren, wie es in der vorgelegten Arbeit geschehen ist.

Die Flugzeugbesatzungen nutzen beim Auftreten von Turbulenz ihre Erfahrungen. Sie suchen während des Fluges in erlaubten Grenzen eine Flughöhe mit geringer Flugunruhe (CAT) auf und meiden Gebiete, wo nach dem Wetterkartenbild starke Turbulenz zu erwarten ist. Die Triebwerke werden bei Eintritt der Luftunruhe gedrosselt, um bei geringer Fahrt die Beanspruchung des Fluggerätes durch die wechselnden Böen herabzusetzen.

9.5 Grenzen für ein vollautomatisches Abrufverfahren meteorologischer Daten

Die Berechnung des Windes setzt eine kritische Behandlung der registrierten Daten voraus. Mit den Instrumenten im Flugzeug werden während Kursänderungen, Steig- oder Sinkflügen keine brauchbaren Meßwerte aufgezeichnet. Großen Einfluß haben dabei die unterschiedlichen Empfindlichkeiten, Einstellzeiten und Dämpfungen bei der Registrierung und Messung der Vektoren Grundgeschwindigkeit und Flugzeugeigengeschwindigkeit.

Eine vollautomatische Auswertung, z.B. in einem bodenseitigem Abrufverfahren, fordert gleiche Empfindlichkeit für die Messung und Registrierung der Daten. Die Dämpfung muß Resonanzen ausschließen. Zur Erfassung mikrostruktureller Vorgänge der Atmosphäre darf das Gerät keine hohen Einstellzeiten besitzen. Bei einer derartigen Auslegung des Instrumentariums gewännen kurzperiodische Schwingungen des Meßträgers großen Einfluß. Das Gerät sollte deshalb auf die Erfassung mesostruktureller Vorgänge ausgerichtet werden und nur die vertikale Beschleunigung des Flugzeuges kleinräumig erfassen. Die letzten einwandfreien Daten vor einer Höhen- oder Kursänderung sollten gespeichert die Werte der Übergangszone ersetzen.

Die Eingriffe des Menschen auf die Bewegungsvorgänge des Flugzeuges schränken die allgemeine Gültigkeit der Ergebnisse ein. Der Einzelfall einer Turbulenz-Registrierung braucht nicht charakteristisch für einen größeren Raum zu sein. Daher ist selbst bei einer objektiv registrierenden Anlage keine einfache statistische Betrachtungsweise und Auswertung gestattet.

10. Schlußbemerkungen

Die vorliegende Arbeit wurde im Jahre 1962 bei der Deutschen Lufthansa mit einer Forschungsgruppe begonnen. Das Projekt fand die Unterstützung des Herrn Bundesminister der Verteidigung. Im Herbst 1963 übernahm die Flugwissenschaftliche Forschungsanstalt München die aufgearbeiteten Registrierungen und erstellten Auswertungen. Die Arbeitsgruppe bei der Deutschen Lufthansa wurde aufgelöst.

Ich möchte es nicht versäumen, allen Herren zu danken, die sich um finanzielle Unterstützung und die Fortführung des Projektes bemühten. Besonderer Dank gebührt Herrn Dr. Tanck für seinen Einsatz zur Weiterbeschäftigung der Gruppe bei der Deutschen Lufthansa. Herrn Professor Dr. Höltje danke ich für die Unterstützung. Der Deutschen Lufthansa habe ich für die Toleranz zu danken, meteorologische Registrierungen und Untersuchungen mit Düsenverkehrsflugzeugen zu dulden. Ich danke ferner allen Mitgliedern der Flugzeugbesatzungen, insbesondere den Herren Flugkapitänen und Navigatoren, die mir auf Flügen wertvolle Hinweise gaben, Erfahrungen mitteilten oder meine Beobachtungen unterstützten. Dem Deutschen Wetterdienst, speziell den Leitern der Flugwetterwarten Frankfurt und Hamburg, Herrn Dr. Regula und Herrn Dr. Wagner sowie den Herren im Seewetteramt Hamburg danke ich für die leihweise Überlassung der analysierten Wetterkarten und Wettermeldungen.

Herrn Professor Dr. Raethjen sage ich meinen aufrichtigen Dank für die Anregung, Beobachtungen aus Verkehrsflugzeugen meteorologisch auszuwerten und für den Rat, meine berufliche Tätigkeit mit der wissenschaftlichen Bearbeitung von Registrierergebnissen meteorologischer Elemente aus dem Tropopausen-Niveau zu beginnen. Ich danke für die Gespräche und für die wertvollen Hinweise bei der Durchführung der Arbeit.

Herzlich bedanke ich mich bei meiner Frau für ihre stetige Hilfe bei der Weiterführung und Fertigstellung der Arbeit.

11. Verzeichnisse

11.1 Quellen

1. Registrierstreifen des Düsenverkehrsflugzeuges D-ABOD der Deutschen Lufthansa aus der Zeit von März 1962 bis Oktober 1963

2. Flugprotokolle der Deutschen Lufthansa

3. Analysen der Flugwetterwarte Frankfurt (Main)

4. Analysen der Flugwetterwarte Hamburg

5. Wetterkarten der 200-mb-Fläche des Zentralamtes des Deutschen Wetterdienstes, Offenbach (Main)

6. Täglicher Wetterbericht des Zentralamtes des Deutschen Wetterdienstes, Offenbach (Main)

7. Wetterkarte des Seewetteramtes Hamburg

8. Analysen des Seewetteramtes Hamburg

9. Tägliche 300-mb-Karten der Nordhemisphäre Institut für Meteorologie und Geophysik der Freien Universität Berlin

10. Daily Series, Synoptik Weather Maps, U.S.Department of Commerce, Weather Bureau

11. Wetterkarten der Flugwetterwarte San Francisco

12. Hell-Fax-Karten des Deutschen Wetterdienstes

13. Wettertelegramme aerologischer Aufstiege aus dem Fernschreibnetz des Deutschen Wetterdienstes

11.2 Literatur

1. Bartels, Julius: Statistical Methods
   Terrestical Magnetism and Atmosph. Electr. 1932

2. Bartels, Julius: Statistische Hintergründe für geophysikalische Synchronisierungsversuche
   Nachr.d.Akad.d.Wissensch. 23, Göttingen 1963

3. Beelitz, Paul: Bearbeitung und Ergebnisse der Radiosondenaufstiege anläßlich des 2. Weltvergleichs von Radiosonden in Payerne (Schweiz) 1956
   Inst.f.Meteorologie u. Geoph. Freie Univers. Berlin, Met.Abh. BdVII/Heft 4

4. Berggren, Roy: On the Accuracy of 500 mb Analysis with Spezial Reference to Numerical Forecasting
   Tellus, August 1957 Vol.9 - 3 Seite 323-340

5. Boeing Airplane Company: Boeing 707 Intercontinental Flight Manual, Model 707-430

6. Boucher, Roland J.; Wexler, R.; Atlas, D; Lhermitte, R.M.: Mesoscale Wind Structure Revealed by Doppler Radar
   Journal of Applied Meteorology, Oct.1965

7. Clarke, R.H.: Turbulence and Detailed Structure of a Subtropical Jet Stream
   Journal of the Atmosph.Sciences, Sep.1966

8. Deutsche Lufthansa: Bordmeßgeräte Boeing 707-430

9. Deutsche Lufthansa: Doppler-Radar mit Flugwegrechner
   T-Bericht Nr.14

10. Deutsche Lufthansa: Turbulenz in großen Höhen
    TL-Bericht Nr.4

11. Deutsche Lufthansa: Überschall-Luftverkehr
    T-Bericht Nr.13

12. Döös, Eaton: Upper Air Analysis over Ocean Areas
    Tellus 9 , Nr.2 Seiten 184 - 194

13. Duensing, Georg: Die Genauigkeit der Windbeobachtungen, welche von Verkehrsflugzeugen der Deutschen Lufthansa in den Monaten Oktober 1959 bis März 1960 über dem nordatlantischen Ozean erflogen wurden (unveröffentlicht)

14. Duensing, Georg: Erfahrungen über Bordregistrierungen von Wind, Temperatur und vertikaler Beschleunigung im Düsenluftverkehr der Deutschen Lufthansa

Bericht Deutsche Lufthansa, Hamburg 1963

15. Endlich, R.M.; Rados, R.M.: The Meteorological Measurements and Field Program of Project Jet Stream from 1956 to 1958

Geophysical Research Papers 64, Oct.1959
U.S.Air Force Cambridge Research Center

16. Essenwanger, Oskar: Probleme der Windstatistik

Meteorologische Rundschau, 1959 Heft 2

17. Essenwanger, O.: Neue Methode der Zerlegung von Häufigkeitsverteilungen in Gauss'sche Normalkurven und ihre Anwendung in der Meteorologie

Berichte Deutscher Wetterdienst Nr.10 , 1954

18. Essenwanger, O.: Tafeln zur Häufigkeitszerlegung mit Anwendungsbeispielen

Berichte Deutscher Wetterdienst Nr.39 , 1957

19. Essenwanger, O.: Zur Häufigkeitsanalyse meteorologischer Beobachtungen

Zeitschrift für Meteorologie 9 , 1955

20. Fett, Walter: Studie zur Typenbildung des Höhenwindfeldes

Meteorologische Abhandlungen XXXIII - 1
Inst.f.Met.u.Geoph.d.Freien Universität Berlin 1963

21. Finger, F.; Harris ; Teveles: Diurnal Variation of Wind, Pressure and Temperature in the Stratosphere

Journal of Applied Meteorology Vol.4 , 1965

22. Flohn, Hermann: Aktuelle Probleme der aerologischen Synoptik

Berichte Deutscher Wetterdienst Nr.51 , 1958

23. Fujita, Tetsuya: Accurate Calibration of Doppler Winds for their Use in the Computation of Mesoscale Wind Fields

Monthly Weather Review, 94-1, Jan. 1966

24. von Hackewitz, Helga: Ein Einblick in die Arbeitsmethoden und Ergebnisse mesometeorologischer Untersuchungen

Meteorologische Abhandlungen LI - 1 , 1964
Inst.f.Met.u.Geoph. Freie Universität Berlin

25. Harris, M.F.; Finger; Teweles: Diurnal Variations of Wind, Pressure and Temperature in the Troposphere and Stratosphere over the Azores
Journal of the Atmosph.Sciences 19, Mar.1962

26. Hering, W.S.; Borden, T.R.: Diurnal Variations in the Summer Wind Field over the Central United States
Journal of the Atmospheric Sciences Jan. 1962

27. Höflich, Otto: Konvektionsfronten und Druckfeldänderungen im jet stream
Hamburger Geophysikalische Einzelschriften Heft 4 , 1961

28. de Jong, H.M.: Geostrophic Approximation in the Upper Air Flow with Application to Aeronavigation and Air Trajektories
Koninklijk Nederlands Meteorologisch Institut Medelingen en Verhandelingen No.74

29. Kao, S.K.; Woods, H.D.: Energy Spectra of Meso-Scale Turbulence Along and Across the Jet Stream
Journal of the Atmospheric Sciences, Vol.21, 1964

30 Koller: Graphische Tafeln zur Beurteilung statistischer Zahlen
Steinkopf, Darmstadt

31 Koppe,H.: Probleme der Höhenmessung in der Luftfahrt
DFL-Bericht Nr.159

32 Krishnamurti: The Subtropical Jet Stream of Winter
Univ. of Chicago, 1959

33. Kuettner, Joachim: The Band Structure of the Atmosphere
Tellus, Vol.11 - 3, Aug.1959

34. Laby, J.E.; Sparrow, J.G.: Wind Studies Using Level Ballons
Journal of Applied Meteorology, Vol.4, Oct.1965

35. Lester, H.C.; Tolefson, H.B.: A Study of Launch-Vehicle Responses to Detailed Characteristics of the Wind Profile
Journal of Applied Meteorology, Vol.3, Oct.1964

36. Linke: Meteorologisches Taschenbuch
Bände 1 - 3 , Leipzig 1953

37. Linder, A.: Statistische Methoden
Birkhäuser, Basel 1960

38. Panofsky, H.A.: Spectra and cross-spectra of velocity components in the mesometeorological range
Quart.J.Meteor. 81 , 1955

39. Prandtl, Ludwig: Strömungslehre
Vieweg , Braunschweig 1965

40. Raethjen, Paul: Einführung in die Physik der Atmosphäre
Teubner , Leipzig 1942

41. Raethjen, Paul: Kurzer Abriß der Meteorologie 1 - 4
Geophys. Einzelschriften, Hamburg

42. Raethjen, Paul: Dynamik der Zyklonen
Geest & Portig , Leipzig 1953

43. Raethjen, Paul: Zur Dynamik des jet stream Gleichgewichtsstörungen im nichtstationären jet stream
Geophys. Einzelschriften, Hamburg 1961, Heft 4

44. Raethjen, Paul: Bemerkungen zur "Clear Air Turbulence"
Beiträge zur Physik der Atmosphäre, 36 ,1963

45. Raethjen, P.: Über die Anwendung der technischen Turbulenzlehre auf die Reibungsschicht der Atmosphäre
Geofisica pura e Applicata, Milano Vol.49, 1961

46. Raethjen, P.: Bemerkungen zum Tangentialdruck des Windes am Erdboden
Zeitschrift für Meteorologie, Berlin 1951

47. Raethjen, P.: Über den "Scale"-Unterschied vertikaler und horizontaler Störbewegungen
Beiträge z. Phys.d.Atmosphäre, Frankfurt 1959

48. Raethjen, P.: Über den zweidimensionalen Impulsaustausch rotierender Systeme
Archiv f.Meteorologie, Geophysik u. Bioklimatologie, Serie A , Bd.3 , 1951

49. Regula, H.; Zimmerschied, W.: Luftfahrt-Meteorologie
Akad. Verlagsgesellschaft, Frankfurt 1959

50. Reiter,E.: Meteorologie der Strahlströme
Springer , Wien 1961

51. Reiter,E.: Die vertikale Struktur des Strahlstromkernes aus Forschungsflügen des Project Jet Stream
Berichte Deutscher Wetterdienst Nr.80, 1962

52. Reiter,E.: Die Feinstruktur der Strahlströme
Colorado State University, Fort Collins, 1961

53. Reiter,E.: Clear Air Turbulence: Solutions and Problems
Soc.of Autom.Eng. U.S.A. 1966

54. Reiter,E.: Experiments in objective upper-wind analysis and forecasting
Colorado State University, Fort Collins, 1964

55. Reiter,E.: Atmospheric Structur and Clear Air Turbulence
Colorado State University, Fort Collins, 1964

56. Reiter,E.: A Case Study of Radioactive Fallout
Journal of Applied Meteorology, Dec.1963

57. Reiter,E.: On the Nature of Clear Air Turbulence
Aerospace Engineering, Vol.21-11, Nov.1962

58. Reiter,E.; Glasser,M.E.; Mahlman,J.D.: The Role of the Tropopause in Stratospheric-Tropospheric Exchange Processes
Colorado State University, Fort Collins, 1967

59. Rudloff, Willy: Der Gradientwind im wandernden Luftdruckfeld
Annalen der Meteorologie, Bd.8 , 1958

60. Rudloff, Willy: Zum Tagesgang tropischer Höhenwinde
Deutscher Wetterdienst, Seewetteramt Hamburg Einzelveröffentlichung Nr.55 , 1966

61. Seilkopf, H.: Mittelräumige atmosphärische Strömungstypen
Annalen der Hydrographie und Maritimen Meteorologie, 1936 , IX , Beiheft

62. Sommerfeld, A.: Theoretische Physik
Mechanik der deformierbaren Medien
Geest & Portig , Leipzig 1957

63. Stone, P.H.: Frontogenesis by Horizontal Wind Deformation Fields
Journal of the Atmospheric Sciences, 23, Sep.1966

64. Sußebach, J.: Der Spotwind
Berliner Wetterkarte, Beilage 22/65, 5.3.1965

65. Sußebach, J.: Windscherungen in der freien Atmosphäre
Inst.f.Meteorologie u.Geophysik d.Freien Universität Berlin, Meteorologische Abh. Bd.LX-3 Berlin 1966

66. Tanck, H.-J.: Meteorologische Schulungshefte für Flugzeugbesatzungen der Deutschen Lufthansa

67. Van der Hoven: Power Spectrum of Horizontal Wind Speed
Journal of Meteorology, Apr.1957

68. Weber, E.: Grundriß der biologischen Statistik
Fischer, Jena 1964

69. Weber, F.: Vorläufige Übersicht über Untersuchungen zur Frage der wolkenfreien Turbulenz
Referat Arbeitskreis 2/4 der Deutschen Gesellschaft für Ortung und Navigation, 5.5.1964

70. Weber, F.: Ergebnisse meteorologischer Messungen mit Verkehrsflugzeugen im Bereich von Höhentief und Höhentrog
Vortrag, Deutsche Meteorologische Gesellschaft Tagung 1966 , München

71. Weber, F.: Monographie der Strahlströme
Forschungsbericht Nr.5 , 1959
Deutsche Forschungsanstalt für Segelflug

72. Weber, F.: Turbulenzmessungen bei Jet-Stream-Wetterlagen
Forschungsbericht Nr.9 , 1961
Flugw.Forschungsanstalt München e.V.

73. Worthley, L.E.: Deviation of Geostrophic Wind from Measured Wind at 500 mb
University of Hawai, Report No.4, Apr.1959

74. Zurmühl, R.: Praktische Mathematik
Springer , Berlin 1957

75. Air Force Surveys in Geophysics No.140
Proceedings of the national symposium on winds for aerospace vehicle design
Vol. I , II

76. Bundesminister für Verkehr, Deutscher Wetterdienst, Bundesanstalt für Flugsicherung, Deutsche Lufthansa: Betriebsanalyse des Düsenluftverkehrs, Reihe Paper-Jet
Frankfurt, Offenbach 1961

77. Rosemount Engineering Company: Total Temperature Probe Model 102 Series
Rosemount Bulletin 7597

In der hier vorliegenden Reihe "Hamburger Geophysikalische Einzelschriften" (bis einschließlich Heft 3 "Geophysikalische Einzelschriften") sind bisher erschienen:

Heft 1 Raethjen, P.
Kurzer Abriß der Meteorologie - dynamisch gesehen,
Wärmehaushalt der Atmosphäre, Hamburg 1950

Heft 2 Raethjen, P.
Frontalzyklogenesis, Hamburg 1954

Heft 3 Raethjen, P.
Kurzer Abriß der Meteorologie - dynamisch gesehen,
Zirkulationen der Atmosphäre, Hamburg 1954

Heft 4 Zur Dynamik des jet stream, Hamburg 1961, in zwei Abhandlungen:

Raethjen, P.
Gleichgewichtsstörungen im nichtstationären jet stream
Disturbance of balance of the nonsteady jet stream

Höflich, O.
Konvektionsfronten und Druckfeldänderungen im jet stream
Convection fronts and development of pressure pattern in the jet stream

Heft 5 Strobach, K.
Ein Beitrag zum Problem der Entstehung und der Wellennatur der mikroseismischen Bodenunruhe, Hamburg 1962

Heft 6 Jeske, H.
Die Ausbreitung elektromagnetischer Wellen im cm- bis m-Band über dem Meer unter besonderer Berücksichtigung der meteorologischen Bedingungen in der maritimen Grenzschicht.
Hamburg 1965

Heft 7 Höflich, O.
Biometeorologische Untersuchungen von Phantomschmerzen,
Hamburg 1966

Heft 8 Raethjen, P., K. Knolle, M. Dunst und G. Fischer
Experimentelle und theoretische Untersuchungen über die atmosphärischen Zirkulationen, den jet stream und die zyklonalen Störungen
Experimental and theoretical research on the atmospheric circulations the jet stream and the cyclonic disturbances, Hamburg 1967

In der Reihe "Berichte des Instituts für Radiometeorologie und Maritime Meteorologie an der Universität Hamburg, Institut der Fraunhofer-Gesellschaft" (Direktor: Prof. Dr. K. Brocks) sind bisher erschienen:

Bericht Nr. 1: Brocks, K.
Meteorologische Untersuchungen der elektromagnetischen Wellenausbreitung im Cm- bis M-Band vor der deutschen Nordseeküste, Hamburg 1958

Bericht Nr. 2: Brocks, K.
Feldstärkeregistrierungen auf Richtfunkstrecken im Cm- bis M-Band über der Nordsee und ihre Beziehungen zu meteorologischen Vorgängen, Hamburg 1959

Bericht Nr. 3: Brocks, K.
Richtfunkstrecken über See - Feldstärkewerte, Hamburg 196o

Bericht Nr. 4: Brocks, K. und H. Jeske
Analyse gleichzeitiger Feldstärkeregistrierungen über der Nordsee auf den Frequenzen 7; 2; o,6; o,2 und o,16 GHz. Hamburg 1961

Bericht Nr. 5: Brocks, K., H. Cassebaum und H. Jeske
Richtfunkstrecken über See - Feldstärkewerte, Teil II, Hamburg 1962

Bericht Nr. 6: Brocks, K., H. Cassebaum, G. Fengler und H. Jeske
Meteorologische Beobachtungen in der Helgoländer Bucht 1956-1961. Hamburg 1963

Bericht Nr. 7: Brocks, K., G. Fengler und H. Jeske
Radiometeorological Papers
Radiometeorologische Vorträge, Hamburg 1963

Bericht Nr. 8: Fengler, G.
Untersuchungen der elektromagnetischen Wellenausbreitung im 500 MHz-Bereich über Land unter besonderer Berücksichtigung der meteorologischen Bedingungen, Hamburg 1964

Bericht Nr. 9: Fengler, G., H. Jeske und G. Stilke
Radiometeorological Papers II
Radiometeorologische Vorträge, Teil II, Hamburg 1964

Bericht Nr. 10: Seehars, H.-D.
Fadinguntersuchungen im Zentimeter- bis Meter-Band auf Richtfunkstrecken über See, Hamburg 1966

Bericht Nr. 11: Kruspe, G.
Untersuchung von Brechungswertschichtungen über See mit einem im Flugzeug installierten Mikrowellenrefraktometer unter besonderer Berücksichtigung meteorologischer Aspekte.
Research on refractive index structures over sea by airborne refractometer measurements with special regard to meteorological aspects.
Hamburg 1966

Bericht Nr. 12: Müller, H.
Messungen der Brechungsindex- und Temperaturschichtung im Bereich von Richtfunkstrecken - Ergebnisse von Flugzeugaufstiegen über der Deutschen Bucht.
Aircraft measurements of refractive index and temperature in the range of radio transhorizon links over sea.
Hamburg 1967

Von den "Technischen Mitteilungen aus dem Institut für Radiometeorologie und Maritime Meteorologie an der Universität Hamburg und dem Meteorologischen Institut der Universität Hamburg" liegen bisher vor:

Nr. 1: Kruspe, G.
Der atmosphärische Brechungsindex und seine Messung mit Mikrowellen-Refraktometern, Hamburg 1965

Nr. 2: Seehars, H.-D.
Die Wellenausbreitung im m- bis cm-Bereich unter besonderer Berücksichtigung troposphärischer Streuprozesse, Hamburg 1966

Walter, H. und W. Gieraths
Über den Aufbau einer Phasenmeßanlage für 9,4 GHz, Hamburg 1966

Nr. 3: Becker, K.-D.
Die Ausbreitung elektromagnetischer Wellen über kugelförmiger Erde in homogener und inhomogener Atmosphäre, Hamburg 1967

Die "Berichte" und "Technischen Mitteilungen" können beim Institut für Radiometeorologie und Maritime Meteorologie an der Universität Hamburg, z.Zt. 2000 Hamburg 13, Binderstr. 22, angefordert werden.

In der vom Institut für Meereskunde der Universität Hamburg (Direktor: Prof. Dr. W. Hansen) herausgegebenen Veröffentlichungsreihe "Mitteilungen des Instituts für Meereskunde der Universität Hamburg" sind bisher erschienen:

Nr. I Proceedings of the Symposium on Mathematical Hydrodynamical Methods of Physical Oceanography, September 1961. Hamburg 1962

Nr. II Technical Report of the Subcommitee on Oceanographic Research. Current Measurements in the Faeroe-Shetland Channel 1961. Tables. Hamburg, März 1964

Nr. III Friedrich, H.
Numerische Berechnungen der allgemeinen Zirkulation im Meere nach einem Differenzenverfahren, vornehmlich für den Atlantischen Ozean. Hamburg, April 1966

Nr. IV Sündermann, J.
Ein Vergleich zwischen der analytischen und der numerischen Berechnung winderzeugter Strömungen und Wasserstände in einem Modellmeer mit Anwendungen auf die Nordsee.
Hamburg 1966

Nr. V Hansen, W.
Die Reproduktion der Bewegungsvorgänge im Meere mit Hilfe hydrodynamisch-numerischer Verfahren.
Hamburg, Dezember 1966

Nr. VI Hansen, W.
Preliminary scheme of deep sea, tidal measurements based on tidal theory. Hamburg, Dezember 1966

Nr. VII Brettschneider, G.
Anwendung des hydrodynamisch-numerischen Verfahrens zur Ermittlung der M_2-Mitschwingungsgezeit der Nordsee.
Hamburg, Januar 1967

Nr. VIII Brettschneider, G.
Modelluntersuchungen der Gezeiten der Nordsee unter Anwendung des hydrodynamisch-numerischen Verfahrens.
Hamburg, Juni 1967

Nr. IX von Trepka, L.
Anwendung des hydrodynamisch-numerischen Verfahrens zur Ermittlung des Schelfeinflusses auf die Gezeiten in Modellkanälen und Modellozeanen. Hamburg, Juni 1967

Nr. X Proceedings of the Symposium on Mathematical-Hydrodynamical Investigations of Physical Processes in the Sea.
Moskau, Mai 1966. Hamburg, Februar 1968

Herstellung: Xerographie G. M. L. Wittenborn Söhne, Hamburg 13

FSC
www.fsc.org
MIX
Papier | Fördert
gute Waldnutzung
FSC® C083411

Kolibri 360 GmbH
Ferdinand-Jühlke-Straße 7
99095 Erfurt, Deutschland
produktsicherheit@kolibri360.de